SPACE
EXPLORATION

One scenario for a future Mars mission, using the Orion capsule as a crew excursion vehicle for exploration at Mars. (Courtesy Lockheed Martin)

SPACE EXPLORATION

PAST, PRESENT, FUTURE

CAROLYN COLLINS PETERSEN

AMBERLEY

To John, Mary, and Mark, who kept me looking up.

First published 2017

Amberley Publishing
The Hill, Stroud
Gloucestershire, GL5 4EP

www.amberley-books.com

British Library Cataloguing in Publication Data.
A catalogue record for this book is available from the British Library.

ISBN 978 1 4456 5603 8 (hardback)
ISBN 978 1 4456 5604 5 (ebook)

Typesetting and Origination by Amberley Publishing.
Printed in the UK.

Contents

Introduction

Imagine you could go to college and get an undergraduate job controlling a mission to another planet while you're working on your degree, or work on a senior thesis analysing data from the Hubble Space Telescope (HST), or spend a semester working with an aerospace contractor, learning the ins and outs of spacecraft design. These are opportunities available to young people today thanks to the Space Age. They were available to me as a graduate student, and this changed my life. As a young child, I often dreamed of being an astronaut or living on the Moon. It just seemed like that was the future. So, when an opportunity to work on a space-based project came up in college, I leapt at the chance and eventually ended up working on an HST instrument team. That experience led me to co-author two books about the telescope's discoveries and advanced my science writing career.

The type of opportunity offered to me exists at many colleges and universities around the world. Today, students continue to control satellites at my alma mater – the University of Colorado at Boulder. They also work on spacecraft design and weather observations and many other scientific endeavours involving space. Around the world, college and high-school students build their own satellites and get them launched to orbit. They're learning science and engineering through hands-on experience, taking their own rightful place as part of our spacefaring species.

The Space Age we live in provides opportunities and technologies that change and enhance our lives. I work on a computer many times more powerful than the first ones used on the space shuttles. Its screen is decorated with scenes of space art, reminding me that the universe is ours to explore. As a long-time science fiction fan, I am often piqued by the disparity between the futuristic series I love to read (and watch on television and in the movies) and the reality of our current space exploration efforts here on Earth. Of course, in science fiction, writers tell stories set inside societies that have long solved the issues we are still battling both within our societies and in our own first steps to space. We've had to create our spacefaring culture one step at a time. I suspect there will always be a disconnect between the space exploration of our dreams and the reality of our efforts in the here and now. The idea of space exploration, and the type of cultures it requires and creates, is something we will continue to pursue for as long as humans look to the sky and wonder about what's out there.

When I first set out to write this book, I was intrigued with the idea of how we will evolve into a spacefaring species. I decided I wanted the book to answer the question, 'What does it take to build a spacefaring civilisation?' for readers who may ask the same question but haven't had a chance to read much about how we came to be in the Space Age. I began to answer that question first by listing the things I thought were important to this goal. Among them are our civilisation's past history and its technological level, the reasons it has for yearning to explore beyond its home planet, the people involved, the companies and institutions working on the resources needed to accomplish flights to space, the education of a spacefaring species, and the future it wants to build for itself.

The history of space exploration is a complex story. It involves talking about the work of men and women with vision and genius and tenacity who wanted to get us to space. They did what it took to accomplish the first flights, build the first space stations, take the first steps on the Moon, send the first robots to Mars, and so on. It is also a story about science and technology invented and applied and pursued at universities, research institutions, space agencies, by governments and private industry.

The story of space exploration doesn't really start with the first rockets built by Robert H. Goddard or the earliest airplanes to fly. It stretches far back into antiquity and the innovations made by people who wanted to travel, to protect their borders, to spread their influence. Their cultures and civilisations laid the framework for the work we do in space today, just as the first farmers learned to plant the food grains we still cultivate today. From the ancient Chinese kite makers to present-day astronauts and private commercial missile builders, the dream to fly out from Earth continues to captivate the imagination. The people who began our thrust into space spent their lives pursuing and solving engineering challenges. Their work introduced the science of flight and influenced aspects of space exploration ranging from the first launches to plans for long-term habitation in space and on the Moon and Mars.

There are also aspects of living in the Space Age we don't immediately associate with it: literature and cultural expectations, for example. The ability of a civilisation to do great things is also affected by its foresight and interests, which are often expressed in its literature. For space exploration that literature is science fiction, and its writers take us on some incredible journeys of the mind.

Who knows where else their vision will take us? Humans have always dreamed about exploration of the unknown, whether it was the next valley, the next country, the next continent, or beyond Earth. We've always imagined what it would be like to fly like a bird, go to space, walk on another world. This is where science fiction comes from – the dreams of exploration. Science fiction has often been called the 'history of the future', with its all-too-believable tales of space exploration and the people involved.

Many readers are familiar with the tales of Captains Kirk and Picard and Janeway and Archer on the *Star Trek* ships *Enterprise* and *Voyager*. Others follow the *Star Wars* universe, or the *Battlestar Galactica* tales. Science fiction readers grew up reading the works of Asimov, Clarke, Heinlein, Silverberg, Bujold, and others. The tales in these books, TV shows and movies place humanity squarely among the stars, exploring the great beyond to search out new worlds to explore. Often, stories are based on actual science and scientists solving problems of space flight, orbital colonies, planetary exploration and flight between the stars. People inspired by such tales have gone on to become scientists themselves, as well as astronauts, engineers, technical experts and science writers. Ask around NASA or the ESA and you're bound to find people who are Trekkies or first got excited about science when reading a novel about trips to the stars.

Many of the technologies first seen in SF have become reality. Got a mobile phone? Had a mammogram? Been diagnosed using modern medical sensors? Used solar power? Navigated using signals from global positioning satellites? If so, you're benefitting from life in a world that science fiction foresaw and space exploration has enabled. Today's technological society depends on both for technology and advancement. Tomorrow's society will as well, even as we dream of a time when warp-drive ships, transporters and sub-space communicators could become regular aspects of everyday life. Space exploration has changed our societies in countless ways, and will continue to do so well into the future.

In the process of writing, this book has evolved into a set of essays aimed at answering that question from before: What does it take to build a spacefaring civilisation? Where it seems appropriate, I brought in viewpoints from the two science fiction writers who first influenced my thinking: Arthur C. Clarke and Robert A. Heinlein. Their visions – and the works of many other writers I've read over the years – helped shape my own future interests in space, and when the first launches and landings took place they prepared me for what to see.

I want to give you a taste of where space flight came from, what it achieved, and where it's going. It's the general story of space exploration, an overview that gives you the big picture. That involves history, so the first two chapters cover the history of flight and the Space Age. They take a look at where today's world of space exploration comes from and where tomorrow's begins, from the development of flight and rockets to the exciting days of the Space Race.

Next we look at the human factors of space flight, from the training of astronauts to the laws that govern our access to and exploitation of space. Then we will learn more about the many space agencies across our globe and the goals they set out to attain on their missions. We will also look at the role of academia and commercial interests in the exploration of space. Both are necessary to teach our future explorers and supply the technology needed to make our space dreams come true.

I've dedicated two chapters to planetary exploration and astronomy, both enabled by the technological advances made possible by the Space Age. This is because the story of space exploration is not simply that of going out to space and stepping onto other worlds; some of our most intriguing missions are robotic and have revealed the surfaces of other planets and taken our gaze out to the most distant reaches of the universe. Finally, we look into the future of space exploration and what it means for our civilisation. Will we become truly spacefaring? And if so, where will we go? Those are the questions that challenge all of us as we look to a future on the Moon or Mars, and beyond.

In a treatise as historically rich as this one, it is inevitable that some topics will only be touched upon, particularly when summarising complex missions and programs. My aim here is to give a taste of this grand, glorious enterprise we call space exploration. This book is just the start. Think of it as an executive summary, a taster to whet your appetite. For those readers who want to delve more deeply, I've created an extensive reference and appendix section with more technical information. I hope this work launches you on your own journey.

My deepest appreciation goes to my cadre of readers: Wayne Boncyk, Jack Dunn, Davin Flateau, Pieter Kallemeyn, Lawrence Klaes and Mike Murray. A very heartfelt thanks goes to Mark C. Petersen, who carefully read the entire manuscript as my 'second pair of eyes'. Special recognition goes to Andy Chaikin and Leonard David for useful suggestions and commentary, and to the ever-helpful members of the Space Hipsters Facebook group. They cheerfully and relentlessly talk about all things 'space' 24/7 and their discussions helped me discern what interests people about the subject. As well, to my editors at Amberley, I thank them for their patience as I explored the world's space programs and their achievements to bring this book to you.

First Steps: From Earth to the Sky

We are all in the gutter, but some of us are looking at the stars.

Oscar Wilde

The way to the stars is open.

Sergei Korolev

Sail away from the safe harbor. Catch the trade winds in your sails. Explore. Dream. Discover.

Mark Twain

The stars! They are the flame of humanity's desire to seek beyond our world. In the darkest night they sparkle in the eye, and speak to the mind of strange and wondrous places, far removed in space and time. You've probably felt the call of the stars on a clear, dark night. The planets and Moon are also there, moving against the velvety backdrop of night, sparking dreams of exploration in all but the most jaded stargazer.

The view for us looks much the same as it did for our distant ancestors, although they didn't have to contend with light pollution as we do today. Their fascination with the sky was as much cultural as it was practical. Theirs was an interest rooted in survival and timekeeping via observations of the Sun, Moon and stars. What they learned nurtured the science of astronomy, sparking an interest not just in studying the stars and their role in the cosmos but in actually leaving Earth and going to space in person. It's safe to say that we descendants of those astronomers now long to explore the universe they first observed.

Humanity's journey to the stars is still in the 'baby steps' stage. For all our dreams of exploration 'out there', people have not ventured all that far into space. You can see where some of us live in Earth orbit, if you know when and where to look. The International Space Station (ISS) is a regular sight in the sky; people often mistake it for just another high-flying aircraft, but that tiny point of light circles the entire planet every ninety minutes. It represents the work of people and space agencies from fifteen countries. Working aboard are scientists and astronauts, carrying out work that can only be done in space. It's mind-boggling to think about that as you watch the station pass overhead – that men and women are actually 'out there', living and working on the final frontier.

The road to creating space stations and planetary probes and launching rocket trips to the Moon is a long one, stretching far back into antiquity. The path to the stars began in ancient wars, and it remains an ironic fact that the machinery of today's peaceful space travel has its roots in flaming arrows and kites armed with gunpowder-laden weapons. Later on, artists, inventors, pilots, engineers, astronauts and cosmonauts turned them into the tools of space exploration.

Storytellers of eons past might not have been able to fly to space, but they could do the next best thing by sharing tales of gods, goddesses, heroes and fantastic creatures who did. Eventually those stories became reality as people invented flying machines. The descendants of those dreamers now travel the world in ultra-modern versions of the first planes. It is only logical that we would next turn our attention to space and devise the means to get there. Humanity is still taking its first tentative steps beyond the planet, with dreams of moving to the Moon, Mars and beyond.

The history of space exploration is the story of flight. To fly through our atmosphere in an airliner or out to space on a rocket we have to overcome obstacles. The first is gravity. It's a force of nature – what keeps us 'feet down' on the planet. Gravity works across great distances to hold together planets, solar systems, the galaxy and the large-scale structure of the universe.

Earth's gravitational pull, which is referred to as 1G or 'Earth gravity', causes objects on (or very near) the planet to fall toward its surface at a rate of 9.8 metres per second squared. If Earth had no gravity, everything on its surface would simply float off into space. Earth would not be able to hold itself together, and its parts would all go flying apart into the void. Of course, that doesn't happen here, and that 1G of gravity is one of the reasons Earth is habitable. In addition to holding us here it also allows the planet to keep its atmosphere, which supports our planet's astonishing diversity of life. Gravity is strongest at the surface and weaker farther away (physicists and astronomers say that it decreases with altitude). Once people get to space, they experience a diminished gravitational pull from our planet.

Getting to space and thus overcoming gravity requires a lot of rocket power. When people ride a rocket to space or send payloads full of supplies to the ISS, the rocket has to achieve escape velocity – that is, 11.186 km per second. That's what it takes to break free of Earth's pull. That's also why when you see a rocket launch first-hand, it's a big deal. It's noisy. The power of a launch creates a pounding sensation in your chest and a feeling of exhilaration in your brain.

A rocket has to supply an incredible amount of power – called 'thrust' – to get off the ground and deliver a payload or astronauts to space. The most power is expended closer to the ground, and as the rocket gets farther up there is less gravity to overcome. Airplanes and jets fight the force of gravity, too, but they're not leaving Earth for space. They can take advantage of Earth's atmosphere to fly through the air from Point A to Point B. They don't need the millions of pounds of thrust that rockets do; they just need enough to get themselves off the ground and up to a comfortable cruising altitude.

It took a long time to reach a point where humans could regularly fly through the air and out to space using rockets. Airframe designers and rocket engineers had to experiment with their inventions, testing them and learning from their mistakes as well as their triumphs. Flight wasn't an overnight success, but eventually people figured out the fundamental forces and applied them to space travel.

To understand the history of space exploration, where it has taken us and where it will transport future astronauts, let's take a quick trip through the history of human flight, and learn about some of the pioneers of aviation and astronautics.

From Kites to Space Rockets

It's not hard to imagine an early philosopher watching a bird soaring through the air and wondering what forces governed its motions. From there, it was a huge leap to figure out ways for humans to do the same thing. The hurdles were technological, but humans found ways to overcome them and achieve something new.

The first attempts to fly things through the air occurred in ancient China. Craftspeople created kites as rudimentary aircraft a few hundred years BCE, during what was called the Warring States Period. They built them in the fanciful shapes of animals, birds, insects, and dragons – ornately decorated with ribbons, whistles and strings – and tossed them to the winds. Not only were kites beautiful and graceful, but useful. They were used to send signals long distances, and carry messages. The Chinese harnessed the power of the wind and the air as tools for communications, and also as weapons of war.

They also created small hot-air balloons – candle-powered floating lanterns – to take advantage of the well-known fact that hot air rises. Start with an amount of air at one temperature and then heat it. That causes the air (gas) to expand inside the balloon. The balloon becomes lighter than the colder air surrounding it, and so it rises. Put enough hot air into a well-engineered airframe consisting of bamboo and paper, and one could attach messages to soldiers in the field – someday, if the craft was strong enough, one could carry a soldier into battle.

The age of kite and balloon flights in ancient China represented just the first steps in a long journey. The idea of moving through the air continued to thread its way through the dreams of those who wanted to fashion a way for humans to fly on their own. Could we fly like birds? Probably not, but given an apparatus of flight perhaps we could do it in some form. That idea so enchanted Leonardo da Vinci, the Italian sculptor, architect, and painter, that he studied birds and insects and made countless drawings of wings that a person could strap on and use to fly. He also created plans for ornithopters, rotorcraft, gliders and the parachute. There's no evidence Leonardo built any of these, but his drawings are a treasury of engineering design. As with most human endeavours, the technology to achieve flight had to be created and perfected.

Riding Balloons Aloft

Aviation made a giant leap upwards when the French brothers Joseph-Michel and Jacques-Etienne Montgolfier built and flew large passenger-carrying balloons in 1783. Their early flights worked under the same principles as the first Chinese efforts: fill a balloon with hot air and let it rise. At first only experimental animals and a few brave men rode balloons. Later, as they learned to control their craft, balloon fanciers took up tourists and soldiers in mad dashes above the countryside. Balloon flight caught on as a stylish fad, with enthusiasts appearing at expositions and other celebrations. Ballooning had arrived.

Hot air can only get a craft so far, however. It cools quickly, which puts a damper on long flights. Keeping a fire going under a balloon made sense during those early attempts, but it was not the safest thing to do. Eventually, scientists discovered hydrogen gas (which is lighter than air) could be used to keep these craft aloft. It wasn't long before balloons were used to transport people and goods, and were even pressed into military use during wartime.

Eventually, the rise of the Industrial Revolution in Great Britain, the United States (US) and across the world played a role in the development of engines and electrical motors to power balloons across great distances. No longer were they subject to the whim of the winds. These powered airframes became a vital method of transportation.

The Rise of the Zeppelins

In 1900, the *Luftschiff Zeppelin* LZ1 made its maiden flight, and helped establish a successful line of airships that soon came to dominate the skies. They were named for Count von Zeppelin, a German noble and general who founded an aircraft company and ran regular air routes between Europe and the US. His aircraft carried gondolas for passengers, crew, mail and other cargo. They were steered through the air using motors and propellers.

Zeppelin wasn't the only one to develop the technology. Airship builders in the US, Spain, France, South America and Britain created, tested and deployed rigid and semi-rigid airframes for a variety of purposes, including transportation of goods and passengers. As with so many other technologies, it wasn't long before the needs of war influenced their development and use. During the First World War, zeppelins were useful for reconnaissance missions. However, they were incredibly vulnerable to weather conditions and incendiary weapons fire, and that really limited where and how they could be deployed. The newly invented airplanes were more manoeuvrable and useful for pinpoint weapons fire and airborne battles.

After the First World War, the development of airships for transportation continued in the US and Germany. Zeppelin and others saw an avenue for bringing passengers across the Atlantic in style and luxury. The enthusiasm for such travel was so great that New York's Empire State Building sported a mooring mast for airships to use (although it never was called into action), presaging what might have become the Great Age of Airship Travel.

The dreams of pampered flight for the well-to-do, comfortably ensconced below the belly of an airship, were quite a sales pitch for the ambitious Count Zeppelin and his company. By the 1930s, Germany was the only country heavily involved in airship development for passenger travel. In 1937, Zeppelin's *Hindenburg* travelled between New Jersey and Germany, carrying passengers and goods back and forth. However, the success was short-lived and Zeppelin's dreams died quickly. On 6 May 1937, the *Hindenburg* burst into flames on approach to the tower in Lakehurst, New Jersey. It was a disaster that lives on in grainy film footage and panicked radio announcements as the airframe burned and collapsed to the ground. That event, which killed thirty-six of ninety-seven people aboard the craft, effectively stopped the use of hydrogen-filled airframes, especially for transporting people.

Since that time, helium-filled balloons, dirigibles and airships have become the norm and are still used today, mainly for short tourism hops in Germany (aboard actual Zeppelins), aerial surveillance and advertising. Anyone attending a sporting event such as a major US football game, for example, has seen the blimps from various companies (Goodyear, MetLife, DirecTV, etc.) hovering overhead, providing aerial views of the action. Interestingly, human flight to the edge of space via balloon is being reinvented by entrepreneurial ventures such as World View Enterprises. Research and development into safer systems for the generation, storage and use of hydrogen as a balloon gas may again result in human-rated lighter-than-air ships containing hydrogen, rather than the much rarer and more expensive helium.

Lifting into the Air on Wings

The balloons of the late nineteenth century were an important step to the skies, but twentieth-century aviators wanted to do more with controlled flight. While Germany and others focused on dirigibles in the pre-First World War world, people were working on creating powered flight through other means. That meant creating winged aircraft that use air pressure for lift. As the air moves faster over the wing, the pressure is less than that of the air moving under the wing. This creates lift.

Otto Lilienthal and Gliding Flight

Today's hang-glider enthusiasts owe a great deal to the efforts of a man who literally gave his life to advance heavier-than-air flight. Otto Lilienthal was a German engineer who was known as the Glider King, and the first person to understand and describe mathematically the laws of aviation. Like Leonardo da Vinci he studied bird flight, and he discovered an advantage birds had: a curved wing.

During his career, Lilienthal built and patented large-winged contraptions that would allow pilots to control their aircraft by shifting their weight during flight. He tested his own ideas and engineering principles in the hills around Berlin. Lilienthal's longest flights covered up to 820 feet, gliding for long periods of time. His work caught world attention when photographers captured his early efforts on film.

Lilienthal enjoyed a great deal of respect from engineers in the US who were also exploring the possibilities of flight. In particular, the Wright brothers were quite intrigued with his experiments, and were inspired by his work to keep working on their own version of manned flight. Otto Lilienthal put his life on the line in the pursuit of advancing aviation. He continued to test new and improved wings, jumping from hills and cliffs to gain longer glide times. Unfortunately, he was severely injured in a glider accident on 9 August 1896, and died a day later after telling his brother that 'sacrifices must be made'.

Powered Flight

It wasn't long before inventors such as the Wright brothers and Gustav Whitehead were attaching engines to airframes in an effort to perfect powered flight. In the early 1900s they, along with European aeronauts Traian Vuia and Jacob Ellehammer, tried various designs. The Wright plane was a three-axis controlled aircraft, which gave a pilot improved stability and control during flight. While the Wright brothers have generally been credited with the first powered flight on 13 December 1903, there have been competing claims over the years by others for making the first flight, claims that remain in dispute to this day. The Wright brothers' successful flights should have been instantly famous, but the two were notoriously press-shy and worked quietly to improve their aircraft. They continued with their experimental flights, and tried to interest the US military in the possibilities of a flying machine. This led to a contract with the Army Air Corps in 1907. They continued testing and perfecting aircraft designs, all the while fighting patent infringements and other legal actions by competitors.

The work of the Wright brothers and others signalled the beginning of a powered race to the skies by others, including the French inventor Louis Bleriot. He made the first powered flight across the English Channel. Soon the skies were buzzing with small aircraft. Aviators sought to fly longer and higher, setting world records as they went. It was the age of Charles Lindberg, Amelia Earhart, Glenn Curtis, Harriet Quimby and many others.

The Wings of War

The pace of aviation advances and inventions continued in the years between the two world wars. Jets made their debut in the early 1940s, and were used during much of the Second World War. The skies over Europe, the Pacific and Japan thrummed with the sounds of Messerschmitts, Douglas Destroyers, Glosters, Lockheed P-38s and P-80s. Large numbers of regular and turbo-charged aircraft, such as the Boeing B-29, Curtiss P-36 and P-40, Fokkers, Handley Page Halifax, Mitsubishi Zero, Tupolev TB-03s and many others also took to the skies to perform reconnaissance missions, bomb targets and deliver warheads to others.

Once the war ended in 1945, development of passenger aircraft resumed, with companies such as Boeing, Hughes, Grumman, McDonnell and Douglas turning out more efficient planes. Today, these same companies and their successors, plus Airbus and others, continue to create ever-larger and more fuel-efficient aircraft. Air travel now occurs aboard jets of various sizes, ranging from wide-body behemoths carrying hundreds of passengers to the small regional jets, commuter aircraft and private jets that transport just a few people. In addition, helicopters, familiar in both military and civilian applications, ferry a handful of passengers at a time.

Of course, the world being what it is, military jets and choppers continue to be used in conflicts for defensive and offensive reasons, among them bombing, strafing, reconnaissance and evacuation. While the world wars enabled a bloody advancement of the art and science of air travel, the Second World War also brought something new to the scene: the further development and deployment of rockets.

It's All Rocket Science

In the Second World War, rockets delivered devastation to their targets, particularly in London. Later generations grew up with the spectre of intercontinental ballistic missiles (ICBMs) carrying the threat of nuclear warheads during the Cold War. More recently, the deployment of Patriot missile batteries, Tomahawk missile systems and the Soviet Scud and Frog missiles during the Middle East conflicts of the late 1990s illustrate the modern practice of placing missile batteries as part of a state's armament systems. Such missiles number in the hundreds of thousands. They're tipped with warheads that can do tremendous damage and remain a continuing political concern. The use of missile systems for war is a topic we shall not pursue in this book. It's enough to know that the same technology used in war is also what takes payloads and people to space.

The Chinese (long known for their expertise in fireworks, a tiny form of rockets) invented the predecessors of today's missiles – large, gunpowder-loaded projectiles dubbed 'fire arrows' – more than 800 years ago. They lent military superiority, even if the earliest ones were little more than arrows with propellant-laden tubes tied together.

The first time Chinese generals used a true rocket in war was in 1232, when they fought the Mongols. In one battle, soldiers fired their solid-propellant arrows at the invaders and lobbed what may have been the first gunpowder grenades. They shook the countryside and blasted the opponents with noise and fire. The enemy Mongols grasped the idea at once and made their own self-propelled weapons. It was the earliest missile crisis in human history.

From there the idea spread across Asia, particularly in China, Korea and Japan. By the mid-thirteenth century, the rocket was a staple of war machinery as far west as Europe. The monk Roger Bacon worked on gunpowder for rockets, while in France rocket tubes were invented, and in Italy the first torpedoes were used against enemy ships. Rockets were here to stay.

How Rockets Work

The scientific principles that describe the actions of rockets were first set down by mathematics genius Sir Isaac Newton in his laws of motion:

1) every object stays at rest or remains in motion in a straight line until it is acted upon by an external force. This is the definition of inertia. If nothing acts on an object that is moving, then it will stay at a constant velocity until something happens to change that. The most ubiquitous external force we deal with here on the planet is gravity; it has the power to shape the flight of an object.

2) The velocity of an object changes when it is acted upon by an external force. This is usually expressed as $F = ma$, where m is the mass of an object and a is its acceleration. Multiply those together, and you get the force (F) needed to change its velocity. In the case of rocket or airplane flight, the motors and engines have to deliver enough force (called *thrust*) to overcome the pull of gravity.

3) For every action or force in nature, there is an equal and opposite reaction. An apple exerts a force on the table, but the table also exerts the same force on the apple. A rocket exerts a force that lifts it off the planet, and Earth exerts a force right back. If you want the rocket to stay on the ground, you don't do anything. If you want it to leave, then you have to overcome gravity (see 2) with enough power to send the rocket upward.

Rockets and airplanes rely on propulsion to move through the air. Any plane flies from Point A to Point B above the ground and uses its engine to provide thrust along its path of forward motion. Its wings work with differences in air pressure to provide *lift* up into the air. The plane also experiences *drag*, which is caused by the air the craft is moving through. Drag is a retarding force that acts on the plane as it moves through the atmosphere. Essentially, the engine has to provide enough thrust and the wings need to provide enough lift to overcome drag. Other parts of the plane, such as its tail and ailerons (which are also called 'flight control surfaces') give the pilot ways to control the craft and adjust its attitude.

A rocket has similar requirements if it's going to fly anywhere. However, unlike a plane, it can travel higher (into orbital or suborbital flights, as needed) and farther. The two main forces that govern a rocket's trip to space are thrust and drag (which is the same as for an airplane). Thrust (the propulsive force) is provided by a rocket motor (usually a jet engine or a series of them) and fuel (liquid or solid) for the engine to consume. The weight (mass) of the rocket is pulled back to Earth by gravity; weight is the product of the rocket's mass multiplied by the pull of Earth's gravity. A rocket produces thrust by burning fuel and turning it into hot gas. Inside the rocket, the temperatures and pressures are very high, and exhaust gases coming out the nozzle of the engine at high speed literally push the rocket forward (or up).

So, for a rocket to leave Earth, its thrust has to be stronger than the drag from the air and the pull of gravity. The body of the rocket has to be strong enough to withstand the many stresses as it lofts itself and its payload into space. A rocket has to be very well engineered to withstand all the stresses and forces it encounters during flight. Rockets designed to work in space operate under the same principles; when it pushes exhaust out, the exhaust pushes back, and that moves the rocket forward along its path.

Staging

Although the Chinese appeared to experiment with multi-stage 'fire arrows', the early rockets developed in the twentieth century – the V-2s and such – were single-stage launch vehicles. They carried all their fuel and motors in one stage. That worked fine for use here on Earth, to carry a bomb from one country to another. However, using single-stage rockets to achieve orbit is very difficult due to the high velocity needed to get the rocket to space. In addition, cost is a significant factor: every pound of payload sent to space – say, to low earth orbit – can represent at least $1,000 USD to get it there. Usually, it's much more – up to $5,500 per pound. Reusable multi-stage rockets (such as those successfully launched and recycled by SpaceX and Blue Origin) could bring that cost down to below $1,000 in the very near future.

A single-stage-to-orbit rocket has long been a dream of designers, but one has never been used to place payloads in orbit. It's far easier to create a rocket with two or more stages – a multi-stage launch vehicle. Each stage has its own propellant chambers and rocket motors that are designed to deliver maximum performance during the stage's 'burn'. Once the propellant runs out, the stage is ejected from the launch vehicle. As each one drops away, it reduces the mass of the launch vehicle during ascent. The rocket essentially gets lighter as it goes up.

The lowest stage – 'first stage' – has the most powerful motors to lift the launch vehicle up through the densest part of the atmosphere. As the rocket gets higher up, it jettisons the first stage and the second stage kicks in. The second-stage motor is smaller but still very powerful. If other stages are needed, they operate in turn until the launch is complete. Multi-stage rockets have been used for years, mostly successfully, although they do offer more points of failure. Even today there can be staging problems at launch that result in pad explosions or occur when the second stage fails at orbital insertion. In general, however, multi-stage rockets are a mature technology and continue to be developed and built to deliver payloads and crews to space.

Rocket Visionaries and Societies

Making rockets work for space travel involved the expertise of thousands of people across two continents for decades. It wasn't easy, especially in the beginning. For one thing, not everyone took these fantastical machines seriously. In his work *Prelude to the Space Age*, writer Frank Winter summarised one of the prevailing attitudes toward rocketry in the early years of the twentieth century:

'Imagine, grown people really thinking that man could fly to the Moon!' Such was an almost universal public reaction to the small groups of enthusiasts in a half-dozen countries who, as early as the 1920s and 30s, believed in the possibility of interplanetary flight.

Still, a few geniuses persisted. People such as Robert Goddard, Konstantin Tsiolkovsky and Hermann Oberth had their early ideas about flight met with scepticism or even outright ridicule, but they kept working. It took many test flights before their ideas took hold among society, the military and government. Interestingly, the easiest sell was to the public, in the form of 'fantastic stories' about space flight, tales that had been around since the late 1800s and were dubbed 'science fiction'.

Science Fiction and the Birth of Space Flight

At the end of the nineteenth century and into the early decades of the twentieth, such writers as Jules Verne, H. G. Wells, Robert A. Heinlein, Arthur C. Clarke and Konstantin Tsiolkovsky described a future where people soared to space and explored other worlds. Theirs were science-laden stories about adventures to distant places. Verne's *From the Earth to the Moon*, for example, caught the public imagination and the eye of more than one future rocket designer. In the US, ideas about life on Mars – piqued by a mistranslation of the Italian word *canale* to 'canals' when describing straight lines visible on its surface – set Mars mania in motion, even among some distinguished astronomers. It didn't take long before writers seized on the idea of Martians to fuel tales of exploration. Wells' *The War of the Worlds* was as much a science fiction story as it was a horror tale, bringing malevolent aliens to prey on an innocent Earth.

Science fiction stories spread the ideas of space travel and exploration and aliens across wide audiences. Later generations of enthusiasts embraced *Star Trek* and *Star Wars* as much as they did the books and stories of the science fiction writers. The genre continues today, pushing visions of the future, based on the growth and evolution of modern space technology. For example, trips to Mars have long been a staple of science fiction, as well as tales of travel through wormholes to other parts of our galaxy.

The Rise of the Rockets

The first research papers and monographs about rockets and space flight began to appear in the 1920s. Rocket societies formed across the United States, Germany and Russia. Enthusiasts (and not a few scientists) worked on perfecting designs and experimenting with fuels and loads. In 1930 the American Interplanetary Society was formed in New York City, largely made up of science fiction writers and engineers. Soon the group was experimenting with different designs, with its first rocket flight occurring on Staten Island in 1933. The AIS became the American Rocket Society in 1934. It merged with the Institute of Aerospace Sciences to become the American Institute of Aeronautics and Astronautics (AIAA) in 1963.

Rocketeers with their eyes on propulsive technology also got together at California Institute of Technology to experiment with designs and propellants. Led by Frank Malina and Theodore von Kármán, a group called the 'Suicide Squad' tested various designs in a desert arroyo near Pasadena. Eventually, the team developed jet-assisted take-off rockets (JATOs) and formed the core of the group that later founded the Jet Propulsion Laboratory.

The German Rocketeers

In Germany, the Verein für Raumschiffahrt (Society for Space Travel) began as an interest group following the publication of Hermann Oberth's *Die Rakete Zu den Planetenraumen* (*The Rocket into Interplanetary Space*) in 1923. It rapidly grew as an amateur rocketry association and had members across Germany developing small rockets. Although there was some military interest in their rockets, the lack of funding hampered the group's efforts. City fathers of Berlin were concerned about rocket experiments being carried out so close to the city, and the group disbanded in 1934. That wasn't the end of their efforts, however; many of the members went on to work developing Germany's V-2 rockets, used toward the end of the Second World War.

Rockets in Russia

Russia had the Group for the Study of Reactive Motion (Gruppa Izucheniya Reaktivnogo), which was based in Moscow. Its members were organised into brigades and worked on projects aimed at solving technical issues with propulsive flight, rocket motors, fuels and the development of both winged and wingless missiles. Among its members were a young technician named Sergei Korolev, who founded his own rocket design bureau and became the head of the Soviet space programme in the 50s and 60s.

British Space Enthusiasts

The British Interplanetary Society (BIS) brought together space flight enthusiasts in that country, people who saw a future in space travel using rockets. Even before the Second World War, members had designed plans for moon rockets. After the war, the group went on to explore ideas for rocket designs, space stations, planetary probes and many other things familiar to space enthusiasts today. Member Sir Arthur C. Clarke (made a Knight Commander of the Most Excellent Order of the British Empire in 1989), who worked on radar systems development during the Second World War, wrote a speculative article about how artificial satellites lofted by rockets could be used to enhance global communications. It was a visionary treatise and cemented his reputation as the 'inventor' of communications satellites. He was also a well-known science fiction writer, with many books and stories to his name. Clarke later went on to become chairman of the BIS, using his position to encourage a new generation of scientists and engineers to consider a future that included space travel. The group continues to publish papers and promote space exploration and commercial opportunities in space, and regularly holds meetings. Today it is a think tank for space development and exploration, with members pursuing research into long-term space flight and interstellar probes.

Making Rockets Work

Before large-scale rocket flight could be achieved, people had to work out both the mathematics governing such trips and the hardware to achieve it. The earliest calculations were made by British mathematician William Moore, who worked out the dynamics of rocket propulsion in his *Treatise on the Motion of Rockets and Essay on Naval Gunnery* in 1813. A few years later, Ukrainian Alexander Dmitrievich Zasyadko worked out the specifics of launches, and created a unit in the Russian Army specialising in rocketry.

Konstantin Tsiolkovsky

Advances in rocketry continued, mainly for military use. In 1903, the Russian scientist Konstantin Tsiolkovsky (1857–1935) speculated on the idea of sending people into space, inspired by science fiction writer Jules Verne and his stories of space travel. That year, he published *The Exploration of Cosmic Space by Means of Reaction Devices*. His calculations for achieving orbit and his designs for rocket craft set the stage for later developments. His rocket equation related the change in velocity for a rocket to the effective exhaust velocity (that is, how fast the rocket goes per unit of fuel it consumes), this being known as 'specific impulse'. It also takes into account the mass of the rocket at the beginning of the launch and its mass when the launch is finished.

Tsiolkovsky has long been considered the father of rocket science and dynamics, and his work was the basis for later achievements by such well-known rocket experts as Sergei Korolev, Valentin Glushko and Hermann Oberth. He is also often cited as the developer of *astronautic theory*, which deals with the physics of navigation in space. To develop that, he carefully considered the types of masses that could be delivered to space, the conditions they would face in orbit, and how both rockets and astronauts would survive in the conditions of low Earth orbit.

In addition to his theoretical work, Tsiolkovsky developed aerodynamics test systems and studied the mechanics of flight, publishing papers on dirigibles and the development of powered airplanes with light fuselages. He stopped working on astronautics prior to the First World War, and spent the post-war years teaching mathematics. He was honoured for his earlier work on astronautics by the newly formed Soviet government, which supplied backing for his continued research. When Konstantin Tsiolkovsky died in 1935 all his papers became the property of the state, and for a while they were a closely guarded secret. Nonetheless, his work influenced a generation of rocket scientists around the world.

Robert H. Goddard

In the US, rocket pioneer Robert H. Goddard (1882–1945) did much of his work near Worcester, Massachusetts. He was a scientist working at Clark University, and his main interest was studying the atmosphere using rockets to loft instruments high enough to take temperature and pressure readings. That drove him to experiment with rockets as a possible payload delivery technology. He had a hard time getting funding to pursue the work, but eventually persuaded the Smithsonian Institution to support his research. As he wrote in his 1919 treatise *A Method of Reaching Extreme Altitudes*, interest led to theory:

> A search for methods of raising recording apparatus beyond the range for sounding balloons (about 20 miles) led the writer to develop a theory of rocket action, in general...taking into account air resistance and gravity. The problem was to determine the minimum initial mass of an idea rocket necessary, in order than on continuous loss of mass [through the burning of propellant], a final mass of one pound would remain at any desire altitude.

His experiments with different rocket configurations and fuel loads led to work with solid propellants, beginning in 1915. He eventually switched over to liquid fuels, which was a more difficult and challenging task, requiring fuel tanks, turbines and combustion chambers that hadn't been fashioned for this kind of work.

On 16 March 1926, Goddard sent a rocket up on a 2.5-minute flight that covered just over 12 metres. This gasoline-powered rocket was the first step toward further developments in rocket flight. Goddard worked on newer and more powerful designs using bigger rockets, solving problems in controlling the angle and attitude of rocket flight. He invented the liquid-fuelled rocket and was the first to attach a de Laval nozzle to accelerate the hot exhaust and produce greater thrust. Goddard also developed a gyroscope system for flight control, a payload compartment for scientific instruments and, eventually, a parachute recovery system to return rocket and payload safely to the ground. He also patented the multi-stage rocket in common use today. His 1919 paper, plus his other investigations into rocket design, are classics in the field.

Goddard's work attracted both good and bad attention. The nascent German rocket groups learned of his research and were inspired to improve their own rockets. However, at home, Goddard's early experiments were criticised in the press as fanciful and overly sensational. That attitude was based mainly on a misunderstanding of the science he was doing. *The New York Times* was especially withering in its criticism, mocking his predictions that rockets might someday be able to circle the Moon and take humans and instruments to other worlds. The paper undertook to school Goddard on the relation of action to reaction, positing that (in the journalist's infinite wisdom) there was a need to have something better than a vacuum for a rocket to react against. It was a sorry piece of science writing, entirely wrong in its premise, compounded by undeserved withering sarcasm.

The *Times* eventually retracted its erroneous article forty-nine years later, the day after three astronauts landed on the Moon on 16 July 1969. It was one of the most understated admissions of error in the history of journalism: 'Further investigation and experimentation have confirmed the findings of Isaac Newton in the 17th Century and it is now definitely established that a rocket can function in a vacuum as well as in an atmosphere. *The Times* regrets the error.'

Much of Goddard's work can be read in the original through the auspices of the Smithsonian Institution Archives, where many of his papers are archived.

Hermann Oberth

Russia and the US were not the only states with scientists working on rocket flight. In Germany, Hermann Oberth (1894–1989) was doing theoretical work on the propulsion needed to send rockets to space. Based on his published treatises, rocket societies sprang up dedicated to creating and perfecting the technology. The Verein für Raumschiffahrt (Society for Space Travel), went on to work on the V-2 rocket design. It eventually became a workhorse of the German Army during parts of the Second World War, particularly in bombing raids against London.

Like Goddard, Oberth was interested in the development of liquid-fuelled multi-stage rockets, particularly for long-range bombardment. He suggested the German War Department begin work on such technology, but the department saw no immediate use for the idea. Oberth learned of Robert H. Goddard's work and began doing his own experiments with liquid-fuelled rockets. Eventually, he was able to convince the War Department to follow suit.

Oberth's most famous publication actually started out as his Ph.D. thesis, but was published as a book in 1923. In *Die Rakete zu den Planetraumen* (*The Rocket into Planetary Space*), he speculates about the future of rocket technology:

> ... it is well proven that, with the present state of science and technology, it is possible to construct vehicles that can attain cosmic velocity and that it is probably possible for humans to ride in these vehicles. However, I do not want to close this discourse without writing about whether there is any prospect that such apparatuses will ever really be built. I do not want to claim that this will happen within the next 10 years, but I would like to show the uses of these apparatuses and what they would cost, in order to come to a conclusion about whether they will ever be built.

He also described clearly peaceful uses for rockets, including the lofting of a space telescope to orbit, and building one on an asteroid. He suggested it could contain instruments sensitive to other wavelengths of light, beyond visible, better studied from space.

Oberth was also quite concerned about the effects space flight would have on humans, whom he fully expected to be included on flights. He coined a term, *Andruck*, to describe the G-forces that humans are subject to in flight, and explored ideas about the physical and psychological effects of prolonged exposure to those forces and to their absence (weightlessness). Clearly, he was thinking far ahead to a time when people would be regularly living and working in space.

He also speculated about how a rocket could be used to go to the Moon, circle it and return to Earth. Yet he was also a pragmatist, noting the expense of sending rockets to space simply for scientific missions – a concern that has always been a hallmark of space exploration. To get around such problems he explored the idea of satellites, although he called them 'moons':

> If we let such large rockets circle Earth, they would provide a small moon there, so to speak. They would also no longer have to be configured for descent. Traffic between them and Earth could be maintained with smaller apparatuses so that these large

rockets (we will call them observation stations) increasingly could be transformed for their intended purpose.

Essentially, Oberth predicted today's International Space Station structure, although that's not what he called it. He thought a station could be made habitable for the scientists he envisaged living and working on his little rocket moons.

Before Oberth's early predictions could come true, however, the Second World War intervened, and he and other rocket engineers were packed off to Peenemunde to work under the direction of Wernher von Braun, who had collaborated with Oberth as a youth. At the rocket works, their job was to further develop the V-2 into a terrifying instrument of destruction. Despite their best efforts, the design took longer to perfect than they expected. When it finally was used (very late in the war), it didn't affect the outcome. Hermann Oberth eventually emigrated to the US and worked for von Braun on the Saturn V project, and eventually published a second book of space travel speculations, called *Menschen Im Weltraum* (*Man into Space*).

Wernher von Braun

As a youth, Wernher von Braun was fascinated with the space travel tales of Jules Verne and Oberth's early speculations about rockets into space. He joined the Verein für Raumschiffahrt and indulged his interest in rocket technology and eventually joined the German Army to work on ballistic missiles. During his years in the army he studied physics, receiving a degree in 1934. He eventually went to work on the V-2 rockets during the Second World War, and then worked on transferring V-2 parts and scientists to the US at the end of the war. Upon his arrival in the US, von Braun worked on rockets for the US Army for fifteen years before going to work for the newly formed National Aeronautics and Space Administration (NASA) in 1960. Eventually, he became director of the Marshall Space Flight Center at the site of the Redstone Arsenal.

From Aeronautics to Astronautics

Not all theoretical work on rockets was limited to propulsion systems, although these were certainly important. While aircraft relied on their power and interaction with the atmosphere, rockets were being designed to move out of the atmosphere and into space. There's a distinct difference between the two, and it's more than a scientific delineation between regions. Recall from above that aircraft need power and air to fly. Principles of aeronautics (which is the science of aircraft design and manufacture) help aircraft designers create the best lifting bodies to fly through the air while preserving fuel economy. Simply put, an aircraft needs a certain air 'thickness' in order to operate well. Once beyond the atmosphere a craft cannot rely on air for lift, and a different set of rules come into play.

The science of astronautics tackles the challenges of flight beyond the atmosphere, and the effects spacecraft face: extreme temperatures, gravitational pull, vacuum, radiation (both from the Sun as well as from particles trapped in Earth's magnetic field) and microgravity environment on mass. A vehicle reacts differently in space than it does lower in the gravity 'well' surrounding Earth. It contains the same mass but it doesn't react the same way, which poses problems for navigation and steering. This applies to human bodies as well, and a significant part of astronautics is learning how the environment affects humans (and how designers can protect astronauts from the worst effects of space travel).

In the first decades of the twentieth century, aeronautics and astronautics were largely the same fields of study. That changed as rocket and spacecraft designers came to realise there were distinct differences between the atmosphere and space. Engineer and mathematician Theodore von Kármán worked on problems related to aeronautics and aerodynamics of both aircraft and spacecraft and the effects of the atmosphere on each. He came to the US from Hungary at the invitation of the California Institute of Technology and took on the directorship of the Guggenheim Aeronautical Laboratory in 1930. A few years later, he and his team formed Aerojet Corporation.

Von Kármán (and others) conducted their research into fluid dynamics and atmospheric effects such as laminar flow and turbulence. His work led to the establishment of the Kármán line, which is considered the boundary between the top of Earth's atmosphere and the beginning of space. It lies 100 kilometres above the surface, and at that altitude the atmosphere is so thin that it does not support aeronautical flight and a rocket (or other craft) would technically be 'in space'. It is a rough boundary, often referred to as 'the edge of space', and is subject to varying interpretations in space law. However, if someone asks 'Where does space begin?', the von Kármán definition is the one to use.

The Rockets of War

While small by today's standards, the V-2s (from the German *Vergeltungswaffe* 2, with the technical name Aggregat 4, or A4) still packed enough punch to deliver devastating blows to huge sections of London (for example) during the war. German rocket scientists had bigger plans: to create rockets that would deliver warheads to targets elsewhere in the world – notably the US. At the end of the war the German rocket scientists dispersed – or, more accurately in some cases, were captured by Allied and Soviet forces and taken to the US and Soviet Union to work on rockets for those countries.

Many of the German rocket scientists ended up working for the US Army at the Redstone Arsenal in Alabama, led by von Braun. The Allied armies also captured V-2 parts, which allowed the US to build up to eighty more missiles. The Soviet Union also captured missile production facilities in Germany and continued to build V-2s after the war.

The V-2 is a good prototype of the rockets that later went on to power the Space Age and send payloads into orbit. It was powered by rocket motors that used a mix of ethanol and water for fuel and liquid oxygen (LOX) as an oxidizer. Many rockets in use today for space exploration are of liquid-propellant design. Their fuel flow is precisely controlled and the engines can be turned on or off depending on what's needed. Other rockets use solid fuels, which are a combination of propellant and oxidizer into a single composite material that looks and feels much like a rubber eraser. It is a simpler fuel to use, but cannot be 'turned off' once it is ignited. The US space shuttles used solid rocket boosters (SRBs) to get the system off the ground and liquid-fuelled engines on the shuttles to complete the boost into space once the SRBs were exhausted and ejected during flight.

The original V-2 was built to carry a bomb payload that would be sent along toward its target after the rocket had risen to 80 kilometres in a vertical launch. There were problems with it, forcing von Braun and his team to test and improve the design and fuel mixtures. The

V-2 was first tested successfully on 3 October 1942. At that point, production of rockets ramped up for use in the war.

The V-2 moved at a speed of 5,632 kilometres per hour and carried a 997-kilogram warhead. With a range of just over 321 kilometres, it could easily reach Britain and Ireland. However, it wasn't used in actual warfare until late in 1944, when more than 3,000 were launched toward enemy targets. Not all hit their mark; some fell into the sea or onto regions between Germany and Britain. However, enough of them reached Britain to cause catastrophic damage in many areas.

Dividing the Rocket Spoils of War: The United States
At the end of the war, Allied forces were sent to capture as many rockets as possible along with any rocket scientists they could find. About half the scientists were captured by the Russians, while von Braun and members of his team went to the US in a secret mission called 'Operation Paperclip'. They brought with them more than 300 rail cars loaded with rocket parts: engines, propellant tanks, fuselages, and other equipment. The scientists, their rockets, and intelligence data about the rockets wound up at the White Sands Proving Grounds in New Mexico in the American south-west. Later the entire operation moved to the Redstone Arsenal in Huntsville, Alabama, and von Braun's team assisted the army in building about eighty V-2 rockets.

Ultimately, von Braun and the team developed the PGM-11 Redstone missile as a successor to the V-2. In subsequent years, von Braun began working at NASA on a newer and bigger rocket – the massive Saturn V, which would eventually take people to the Moon.

Today NASA uses a variety of next-generation rockets, including Atlas and Delta Heavy rockets, and other launchers. All are multi-stage, using big boosters to get the rockets off the ground and then switching to additional stages as needed to boost payloads to orbit or further on to other targets. They use guidance and monitoring systems to help keep the rocket on track and for final orbital determination (as needed).

Dividing the Spoils of War: Soviet and Russian Rocketry
The US wasn't the only country to reap the harvest of V-2s at the end of the Second World War. The Soviet Union managed to capture a collection of the missiles and some of the surviving German rocket

engineers to work on them. Ultimately, they used the basic design of the German V-2 to create newer, more powerful versions for use as short-range ballistic missiles under the name R-2. Later, during the Cold War, the R-5, also based on the V-2 design, was used as the Soviet Union's warhead delivery system for nuclear weapons. Other missiles, including the R-7 ICBM, were designed for use in space flight. The R-7 Semyorka evolved into the Vostok rocket, which lofted Sputnik 1 to space in 1957. The Vostok rocket served the Soviet Union throughout the 1960s during its own attempts to win the space race.

In 1966, the Soviets began using the Soyuz rocket, which was a successor to Vostok. This design worked so well that it has become the most-used space launcher in the world, employed in more than 1,700 missions. Soyuz designs are still used today, as well as the Soyuz crew capsule, mainly for delivering crew and supplies to the International Space Station.

Research Rockets for Science and Reconnaissance

Long before the idea of sending rockets to space became a reality, scientists had begun using them to loft instruments through Earth's upper atmosphere. Research rockets (often called sounding rockets) and their payloads were ideal to sample the temperatures, pressures, gas mixes and other characteristics of the atmosphere up through the ionosphere. The US launched its first sounding rockets in 1945, as part of a project between the Jet Propulsion Laboratory and the US Army. They were a major part of the research done during the 1957 International Geophysical Year. Some sounding rockets also carried instruments sensitive to ultraviolet, x-ray and gamma radiation. Whole collections of such rockets were developed, with names such as Aerobee, Astrobee, the Nike and Terrier, Malemute, and the Canadian Black Brant. They could carry extensive payloads, including cameras and test instruments for satellites.

Sounding rockets are still used for atmospheric research. They can be built and equipped relatively quickly, and they can be used for a variety of applications such as studying the upper atmosphere during periods when high solar activity is expected to buffet the ionosphere. Astronomers also use sounding rockets to study x-ray and ultraviolet sources in the sky over short periods of time, and many research groups continue to use them for microgravity research.

China Enters the Space Age

While China may have been the first to develop rocket technology for its ancient wars, it wasn't until the late 1950s that the country began investigating space exploration. As with the US, it took the launch of Sputnik 1 to focus China's attention on becoming a space power.

Part of this was (as it has been for other countries involved in space efforts) a matter of national pride. However, there was also a defensive element. China felt that it too needed the protection of ballistic missiles. Its first missile was essentially a version of the Soviet R-2 short-range ballistic missile system. At that time, China and the Soviet Union cooperated on jointly developing rocket technologies.

The first Chinese ballistic missiles were built and flown in the late 1950s, and China continued to develop more powerful short- and long-range missiles for military use. In the mid-1960s, however, China turned its attention to space. Even something as 'simple' as lofting a satellite into orbit required the development of heavy lift capability, which led to the development of the Long March series of rockets (also known as the CZ series). The Long March rockets continue to be used today.

Europe's Rockets

In the early 1970s, France (and subsequently the European Space Agency, or ESA) proposed to develop and build the Ariane rocket system as a way to get commercial satellites to space for European interests. Although conceived of as a cost-effective way to deliver payloads, Ariane 1 was limited in what it could deliver. Eventually, with more complex and heavier satellites and other payloads, Arianespace (the company that now builds the rockets) developed the Ariane 2 and 3 rockets. Ariane 5 is the current iteration, developed jointly with Airbus Space and Defence, as well as the ESA, the Centre d'Etudes Spatiales and other contractors and subcontractors. Ariane 5 has a liquid-fuel main rocket flanked by two solid-rocket boosters. Its second stage can be ignited multiple times for proper guidance and orbital insertion.

India

India has been developing its fleet of launch vehicles since the 1960s. The country started out (as the US and others did) by sending sounding rockets to the upper atmosphere. Those were followed by the Satellite Launch Vehicle (for lighter payloads) and the Augmented Satellite Launch Vehicle. The current Indian launch capability includes the Polar Satellite Launch Vehicle (PSLV) and the Geosynchronous Satellite Launch Vehicle (GSLV). These use solid rocket boosters and liquid-fuelled engines to get to orbit. India developed some of its technology itself, but has also used trade agreements with Russia to obtain the necessary components.

Japan

Japan is a spacefaring nation and has developed its own series of launch vehicles. Like other countries, Japan became interested in launching its own communication and data satellites, starting with the launch of the Ōsumi satellite aboard the Lambda-4S rocket. Today, JAXA (the Japanese Aerospace Exploration Agency), uses the H-IIA rocket family to send satellites to space and the multi-stage Epsilon rocket to send spacecraft on lunar and planetary explorations, as well as out to small bodies in the solar system.

The Age of Rocketry Continues

In today's age of both government-based and commercial space access (discussed more fully in chapter 6), the cost of space flight has led to the development of more cost-effective, lightweight and reusable rockets. Companies in the US, such as SpaceX and Blue Origin, as well as such commercial entities as Rocket Lab (a New Zealand company that also works in the US) are changing the face of access to low Earth orbit, the Moon and beyond. Delta, Atlas, Ariane, Long March, Soyuz and other rockets continue to deliver payloads to orbit. (There is a complete list of space-based rockets and the launch-capable countries that use them in Appendix A.)

The US shuttle programme, the Space Transportation System (or STS), was a hybrid design of a space plane powered to orbit by solid rocket boosters and rocket engines. It delivered crews and a huge variety of payloads to space (including parts for the International Space Station) from April 1981 through July 2011. It required such heavy lift capability because it could be outfitted to carry up to 23,000 kilograms of payload to space, plus as many as eight astronauts. It was built to launch like a rocket but return to Earth like an aircraft.

The first shuttle, called *Enterprise*, was used for test flights. It never flew in space. Five space shuttles, *Atlantis*, *Challenger*, *Columbia*, *Discovery* and *Endeavour*, had 133 successful launches during the thirty-year programme, and suffered two catastrophic failures. As the fleet aged, the decision was made to stop the programme. The development of capsule-based access to space (similar to Apollo missions) is currently part of NASA's long-term strategy with the Orion Multi-Purpose Crew Vehicle. The European Space Agency is cooperating with NASA on the Orion, which (as of this writing) should fly within the next few years.

The idea of space-plane-type access is not dead. The Dream Chaser Cargo System shuttle-like ship, built by Sierra Nevada Corporation Space Systems, is in advanced development and will rise to space atop an Atlas V, Ariane 5, or Falcon Heavy rocket rather than use the STS's combination of solid- and liquid-fuelled rockets.

Rockets are an integral part of space exploration now and well into the future, regardless of the shape of the crew 'vehicles' they may be lofting to orbit and beyond. While exploration goals evolve to include asteroid missions, lunar exploration and trips to Mars, the ancient technology of the 'fire arrow' stays with us as the most reliable way to get off the planet. The rockets of the future, while perhaps utilising nuclear, ion or laser propulsion, will still be doing the heavy lifting when it comes to space exploration.

From Dreams of Flight to Dreams of Space

For people who lived through the early days of the Space Age in the 1960s, the invention of space capsules and rockets, and the subsequent astronaut trips to space all seemed to burst upon the scene. It's easy to forget that flight didn't begin with the wings and rocket motors we're so familiar with today. The earliest years of the twentieth century saw the invention of the airplane. The First World War saw the first use of aircraft in combat roles in the skies over Europe. The Second World War introduced the use of rockets to warfare. At the end of that conflict, the captured rockets (and the scientists who developed them) made their way to the Soviet Union and the US, which sowed the earliest seeds for eventual space travel.

The Cold War, between the US and the Soviet Union, offered an ever-present spectre of actual conflict between the two superpowers using rockets to deliver warheads. Children in the US learned to 'stand, duck, and cover' in case of a nuclear missile attack. The same precautions were taught in the Soviet Union. At the height of the Cuban Missile Crisis, the world followed for thirteen days as both superpowers grappled over the threat of weapons poised to hit North America from the island of Cuba (a Soviet client state). The proliferation of nuclear weapons bristling on the tips of ICBMs provided a backdrop of world tension throughout the twentieth century, and not just between the two superpowers. Other countries (such as China) built their own arsenals, too. Rocket-guided nuclear weapons (and conventional ones) are still deployed today, silent sentinels awaiting only the push of a button to arc across Earth's skies and deliver their destructive payloads.

The heady and stressful days of the late 1950s and the decade that followed saw Russia and the US competing to get to the Moon first, turning the use of rockets to peaceful means. It all occurred against a backdrop of the Vietnam War, which simmered for years before the US entered into what really turned out to be a quagmire. It was also a proxy fight against the Soviet Union and China on behalf of South Vietnam against North Vietnam.

Social conventions were changing throughout the decade of the 1960s, sending out political, religious, sexual and racial ripples still felt today. This was the socio-political-military tapestry against which the scientists and technicians engaged in the Space Race carried out their work.

By the time the first men landed on the Moon in 1969, more countries were getting interested in space exploration. The US and Soviet Union were creating orbiting laboratories. Engineers and designers began drawing up plans for space shuttles and long-term planetary exploration missions. Decades of space flight have brought space exploration to a point where return trips to the Moon and lunar mines are the new goals. Today we look with longing eyes at a return to the Moon, a trip to an asteroid and, eventually, a human mission to Mars. However, the first step was to the Moon, and that is what we'll examine next.

2

The Space Age: Steps to Orbit and Beyond

Nothing will stop us. The road to the stars is steep and dangerous. But we're not afraid ... Space flights can't be stopped. This isn't the work of one man or even a group of men. It is a historical process which mankind is carrying out in accordance with the natural laws of human development.

Yuri Gagarin

The limits of the possible can only be defined by going beyond them into the impossible.

Arthur C. Clarke

The exploration of space will go ahead, whether we join in it or not, and it is one of the great adventures of all time, and no nation which expects to be the leader of other nations can expect to stay behind in the race for space.

John F. Kennedy

There was a time, not all that long ago, when flight was considered impossible, when leaving Earth was considered a wild fantasy that would never happen. Our history is filled with people making pronouncements about what is and isn't possible, and not just in space flight. In 1985, computer experts opined in a column in the magazine *Infoworld* that 20 megabytes of storage was all anybody with a desktop computer could ever need. That looks ridiculously short-sighted today, when even the cheapest smartphone has many

40

times that amount of storage space. Now people drive computer-enabled cars powered (wholly or in part) by electricity. In some places on roads that are lit from beneath. This is a big change from the first cars, made to go just a few kilometres per hour, based on the rate at which a team of horses could move. People never thought that we'd want to go faster than horses – that was the benchmark of speed.

Times change, technology grows and evolves, and each year we surpass the expectations of people who failed to predict the scale of progress in technology. That's why reading history can be so interesting and amusing. It shows us ideas that were once wild and crazy, and which subsequently changed our lives. Certainly space exploration was one of those ideas that was scoffed at, but not for long.

From Imagination to Reality

Our views of exploration are influenced (for good or bad) as much by science fiction and popular culture as they have been informed by actual space travel accomplishments. Back in the 1920s, the space opera hero Buck Rogers made his first appearance in the novella *Armageddon 2419 A.D.* He was an instant hit and inspired many a child to dream about adventures in space. His name became synonymous with futuristic technology and civilisations on other worlds. The term 'Buck Rogers' has often been used to refer to the kind of hero who could battle against space aliens and evildoers while simultaneously winning the heart of the fair heroine. Sure, it's a cliché, but that is often the role of ideas and icons in pop culture. In the case of Buck Rogers, his rise to fame coincided with that of the Golden Age of Science Fiction in the 1920s.

The reality of space exploration as it unfolded hasn't been quite like Rogers's creator imagined. Nor has it been exactly the way other science fiction writers foresaw the future. Heinlein, for example, in *Rocket Ship Galileo* wrote about three young boys taking a flight to the Moon in the 1940s, when such a feat hadn't occurred in real life. However, it did offer an insight into a society where kids could climb on board a rocket to the Moon for a field trip and parents felt safe in letting them do it.

Sir Arthur C. Clarke's first story, called *Rescue Party*, assumed humans were so advanced that their spacefaring took the form of a race away from danger. It focused on aliens coming to rescue Earthlings from an exploding Sun, only to find that humans had left Earth on multi-generational starships. Nowadays, there are people who think aliens really do exist and insist we've been visited by 'them'. However, that has little to do with the reality of space exploration, other than through the Search for Extraterrestrial Intelligence at California's SETI Institute. In today's view, the search for aliens is something *we* undertake, using powerful telescopes to study distant planets around other stars. No one has found life 'out there' yet, but it will happen.

Clarke was also a visionary who introduced the idea of geostationary communications satellites in 1945. In his treatise *The Space-Station: Its Radio Applications*, he wrote about practical aspects of a space station (including as a refuelling stop for ships leaving the planet):

> However, there is at least one purpose for which the station is ideally suited and indeed has no practical alternative. This is the provision of world-wide ultra-high-frequency radio services, including television.

Clarke goes on to describe how a chain of stations could be used not just to relay communications and data from space to Earth, but for ships at sea to communicate with land and for airliners to contact air traffic control. It was a remarkably prescient forecast of the telecommunications and sensing satellites we rely on today. They were among the first payloads launched to space in the 1960s, and today our planet is encircled by a ring of them relaying our words, texts, videos and emails around the world. His idea of instant communication resonated with readers, although he remarked often that the invention and deployment of communications satellites happened faster than he expected. Based on what science fiction and future 'forecasts' like Clarke's predicted, it's small wonder that young people coming of age in the 1960s flocked to become scientists, engineers, pilots, astronauts and future science fiction writers as the world's space agencies were taking their first steps beyond Earth.

What the activities in the first decades of the Space Age (beginning with Sputnik 1 in 1957) enabled was a solid grounding in flying satellites and space capsules to low Earth orbit. While the goal was the Moon, that period was also the start of planetary exploration, space stations, space shuttles and orbital astronomy. Today, agencies around the world continue the Space Age, building upon the work of the early space pioneers. They're sending new generations of rockets and spacecraft to study the planets and the universe, and ferrying human explorers to space.

Moving into the Space Age

The Space Age began not with a powerful rocket blasting off to orbit, but from a scientist's suggestion that researchers around the world coordinate observations of geophysical phenomena. That idea – to study Earth systems remotely – blossomed into the International Geophysical Year (IGY), when scientists spread out across the planet to observe and chart activities and changes in Earth's atmosphere and oceans. It began on 1 July 1957 and lasted until 31 December 1958. To study the upper atmosphere before that, scientists used sounding rockets to carry instruments to give them data on the geocorona (the glow of far-ultraviolet light from the Sun that reflects off of a cloud of neutral hydrogen surrounding our planet), the chemistry of the upper atmosphere, and the measurement of atmospheric electrical currents and aurorae. The first sounding rocket used for atmospheric studies was launched in 1945, and more than 300 were launched from around the world during the IGY.

The scientific payloads for these rockets included instruments sensitive to various wavelengths of light, spectroscopes to dissect that light, instruments to detect cosmic rays, and much more. In addition, upper-atmosphere balloons (radiosondes) were deployed with instrument packages that measured the temperatures and densities of the various levels of Earth's atmosphere. The IGY was a global effort, encompassing sixty-seven countries and their scientists (including Taiwan but not the People's Republic of China). Most of the scientific studies and investigations were done on Earth's surface and missions to space were planned and announced by the United States and the Soviet Union. However, other countries, including, China, Great Britain, France, Germany, Greece, Japan and Spain, were interested in the applications of space technology, and several were working on their own rocket programs. (We will look at the world's space agencies in more detail in chapter 4.)

The stage was set for *somebody* to start the actual exploration of space. That's what the Soviets did when they launched Sputnik 1 on 4 October 1957. The simple 'beep' signal it sent out told the world, 'Here I am!' Of course, to Americans, it was also a message from the Soviets saying, 'We beat you in to space.' Indeed, it provided the incentive to get the US space programme off the ground. The challenge was to see who could achieve more in space, and it quickly became a space race between the two countries.

While countries like Great Britain and China were also interested in space exploration, as aforementioned, the 'battle' between the two superpowers of the US and the Soviet Union was the pre-eminent effort. The two countries were locked not just in a race to space, but in a 'cold war' that had as one of its goals the domination of space, now the high frontier.

The Vietnam War was ramping up all the while, and the US was fully engaged in it by the mid-1960s. It was very nearly a proxy war with the Soviet Union, and there were enough 'missile scares' through the early 1960s to warrant both countries (and the world) warning citizens about the possibilities of nuclear war.

Against that backdrop, the Space Race was another proxy for conflict between the two countries. Both ramped up their space agencies to meet the challenges of beating the other to space. It mattered who got to space first. It mattered who got to the Moon first. The race consumed a huge amount of effort and money as each country battled to send astronauts to space, boost prestige and *win* the Space Race. In 1969, NASA reported the cost of US missions from the Mercury shots through to the first Moon landing at more than $23 billion. The Soviets are estimated to have spent the equivalent of nearly $10 billion on their Moon missions alone.

The advances of the Space Race developed at breakneck speed. Whole industries, technologies and laws had to be invented and perfected. While the US started out slowly, with many issues getting its rockets to launch perfectly, it eventually crafted a unified plan to get to the Moon, with verifiable goals and benchmarks. The Soviets, while the first to space, eventually fell behind the US, which ultimately 'won' the race to the Moon.

In the larger picture, the actions taken by both countries set the stage for decades of planetary exploration, the development of the Mir and Skylab space stations, the more recent building and deployment of the International Space Station, the use of shuttles (the space shuttle in the US and the Soviet *Buran*), and the building and deployment of the International Space Station in the decades that followed.

Building the Infrastructure of Space: NASA and the Soviet Space Agency

For many years, technology and aeronautics in the US came under the aegis of the National Advisory Committee for Aeronautics (NACA). It was formed in 1915 as an independent agency tasked to coordinate aeronautical research in the US. The first chairman was Brigadier General George Scriven, chief of the Army Signal Corps. He presided over a committee of twelve advisors taken from government, industry and military sectors. At first this committee met a few times each year to discuss new efforts in aeronautics, but as flight technology grew NACA's role expanded.

At the end of the Second World War, NACA pivoted its attention toward designs for advanced supersonic aircraft, and eventually for space travel. The Bell X-1, an experimental rocket plane developed under NACA's direction by Bell Aircraft, was made famous by test pilot Captain Charles 'Chuck' Yeager, who was the first human to break the sound barrier during a test flight in 1947. The design of the X-1 radically changed the look and feel of high-speed flight, and eventually this plane and others in the series helped lead to the development of technologies for space flight.

The launch of Sputnik 1 saw NACA pushed squarely into the Space Race. Politically the effort was guided by military development, but the US turned space flight into a largely civilian enterprise and President Dwight D. Eisenhower ordered the formation of the National Aeronautics and Space Administration (NASA) on 1 October 1958. The new agency absorbed test units and science labs from the Army – including the Redstone Arsenal in Huntsville, Alabama, where Wernher von Braun and his team had been working to develop later advances to the V-2 rockets they'd built in Germany for use in the Second World War. The facilities and personnel at Jet Propulsion Laboratory (JPL) in Pasadena, California were also pulled into NASA. The stakes were high and the agency needed every bit of expertise it could get.

On the other side of the world, the Soviet space programme was primarily a classified military project designed to further that country's strategic aims in space. It began in the 1930s as a rocket development effort under the direction of Sergey Korolev, who went on to become the chief designer of rockets used in the space effort. While the country had benefitted from the advanced work of Konstantin Tsiolkovsky as well as Korolev's ideas, the Soviets also had some rockets and rocket scientists captured from the Germans at the end of the Second World War.

Rocket test facilities were built at Tyuratam (now known as Baikonur Cosmodrome, and colloquially as 'Star City'), where the next generations of rockets based on the V-2 and other designs were improved and tested. The fledgling space programme was important to the country, but most of the military's funding for rocket tests went to the ICBM rockets under the control of the Soviet Strategic Rocket Forces (created in 1959 to manage attacks on enemy missiles).

As with many aspects of government and culture in the Soviet Union in the 1950s, the development of a space programme was part of a larger set of directives called the Five-Year Plans. They outlined both political and economic growth, and particularly in the 1950s and 1960s included space exploration as a large part of the country's economy.

The Soviet Union surprised the world when it launched Sputnik 1 during the International Geophysical Year. That single event, managed by Korolev (although his name was not widely known at the time for security reasons), conferred 'bragging rights' on the Soviets, handing them a huge propaganda boost against the US. From then on, until the founding of the Russian Space Agency in 1992 (later to become Roscosmos, discussed in more detail in chapter 4), the Soviet space efforts were directed from a series of technical and research centres within the union's military and political structure.

Sputnik Starts the Race

The Sputnik mission should not have come as quite the surprise it was. In his landmark book *Red Star in Orbit*, historian James Oberg noted that all the signs were there:

Suspicions that the Soviets were about to embark on some spectacular space venture had been aroused even before [a scientific conference in Washington, D.C.], ever since they claimed that, like the Americans, they, too were preparing to launch scientific satellites as part of the International Geophysical Year. Two months before, they had announced the successful flight of an intercontinental ballistic missile, an ideal booster rocket for such a space probe; only days before they had quietly released the radio frequencies at which their sputnik (Russian for 'traveller') would soon be transmitting.

That first Soviet ICBM flight was made by the R7 Semyorka rocket on 21 August 1957, and it propelled the world into a new reality. The Soviet space programme, steeped as it was in secrecy and conducted under the auspices of the military, was something of an enigma to the rest of the world. Very little was said about its work, and any public pronouncements came only after goals were achieved. Few details about Soviet rocket designs were available, and satellites were mentioned only in vaguest terms (for fear of releasing too much information to the West). The same was true later on for information about space capsules, and even material about the cosmonauts who flew Soviet missions was not immediately available to the world at large. The leadership classified nearly everything about the country's efforts in space, largely because these activities were under military control. Not even Soviet citizens knew what was happening until official pronouncements were made.

The Western intelligence networks had a bit more information, but not enough to predict exactly when a Soviet launch would take place. All in all, the shroud of secrecy hiding the Soviet space programme worked pretty well, which explains why people didn't expect something as flashy as Sputnik 1 to happen as soon as it did.

The launch of the world's first artificial satellite was not an overnight project. Throughout 1956 and 1957, technicians developed the guts of the satellite and its radio transmitter. A rocket from the Soviet R-7 family was chosen to deliver Sputnik to orbit. As with the earlier R7 ICBM launch, the Soviets did drop a few hints that the project was about to happen. Just a few months before the launch, the *Komsomolskaya Pravda* newspaper published an article by the President of the Academy of Sciences Alexander Nesmayenov, making glowing promises about the country's upcoming forays to space and a satellite launch within the year.

The whole Sputnik project and a chance to be the first to launch something to space must have been very exciting for the Soviets involved with the programme. Here they were about to steal a march on the Americans. The payoff of years of development was about to happen. About two weeks before the launch, chief designer Korolev spoke out about what was about to come, in a dedication to the rocket pioneer Konstantin Tsiolkovsky. His words, which should have rung warning bells to the rest of the world and provoked pride in the average Soviet citizen (had he or she known about them), seemed to be taken more as hyperbole than a prediction. 'In the near future, for scientific purposes, USSR and USA will conduct first test launches of the artificial satellites of the Earth,' he said.

On the evening of 4 October 1957, the prediction came true. The carefully constructed and tested rocket with its precious payload was ready to go from the launch pad at Tyuratam. Korolev, along with other officials, watched from a control room in a nearby bunker through the final pre-flight preparations. This was the first time anything this ambitious had been done and the pride of the Soviet Union was on the line. Then a technician pushed a button and the long, silvery rocket rose into the cold night air. It was 22.28.34 Moscow time, and they had just pulled off the most amazing accomplishment of the century.

Many people around the world still remember where they were and how they felt when the news of the Sputnik 1 launch broke. Families gathered around the radio (or the still-new TV) to find out more. Fear was certainly a part of the US reaction, although many were also quite interested in the facts of space travel, too. Politically it was almost a PR disaster for the US, and politicians excoriated President Eisenhower for America's 'loss of primacy' to the Soviets in space.

That 83.6 kg satellite, the size of a beach ball, spurred the US to up its space game, for both political and military reasons. It galvanised the science community, and immediately entered the public consciousness. Within a few hours, radio stations around the world were re-broadcasting the Sputnik signal and the name 'Sputnik' was heard everywhere. Suddenly, everyone was interested in space.

The newly formed NASA rushed to develop rockets and satellites to put the country back on par with the Soviets. Eisenhower worked to pour money into the effort, understanding only too well the loss of face that the country would endure if the challenge to reach space went unanswered. Not only was the US space programme put on the fast track, but development of military weapons programs accelerated too. This was largely due to a fear that the Soviet Union could soon claim primacy in that arena, too. As profound as the event was for the Soviet Union, it was equally if not *more* of a game changer for the Americans, particularly when the Soviets followed up with Sputnik 2 on 3 November 1957. Even so, the US 'played catch-up' to the Soviets until the mid-1960s.

The USA Responds

There were several reasons why the Soviet Union got a satellite launched first. In America, there were two groups in contention for funding to develop and launch satellites. One, funded by the Navy, developed the Vanguard launch vehicle; the other, funded by the Army, worked on the Juno-1, based on the Jupiter-C sounding rockets. That effort was led by Wernher von Braun. A third group, led by the Air Force, proposed the Atlas vehicle, which hadn't yet been built.

The Vanguard was ultimately chosen for the first US attempt at an orbital launch, for political rather than technical reasons. Jupiter was tied intimately to the Redstone ballistic missile programme, and so was regarded as the 'military option'. Vanguard, despite its Navy association, was seen as the more 'civilian' approach since part of its technology was derived from the Aerobee sounding rocket effort (which had been responsible for a decade of suborbital science by that point), and that appealed to President Dwight Eisenhower. They were also arguably slightly ahead of von Braun's Juno effort, based on government assessments of technology maturity for both programs.

Vanguard ultimately failed, twice, with real payloads attached. Both exploded after launch. Eventually, however, NASA managed a successful launch of Explorer 1, the first US artificial satellite, atop a Juno 1 (Jupiter C) rocket on 31 January 1958. That was followed by Vanguard 1 on 17 March 1958, which was originally developed by the Navy Research Laboratory. Explorer 1 was built to detect cosmic rays and micrometeorite impacts. It was the first mission to detect the Van Allen radiation belts (named after James Van Allen, who was the developer of the instrument package.

Vanguard 1 was the first solar-powered satellite and was built to test the environment in Earth orbit. It has been dead for years, but remains in Earth orbit.

Forging Ahead with Human Flight

At the same time the Soviets were working on their next missions, the US built nine Vanguard rockets to take satellites to space. Six of those attempts failed at launch, but three were successful. Against this backdrop of American failures, the Soviets prepared for their next extravaganza: a man in space.

The Soviet space establishment prepared for missions that would allow them to practice with the equipment and techniques to send people to space. They started by sending experimental animals, including Laika, the first dog in space (who died in orbit). After a series of such tests over three years, the stage was set to send a human. The crew capsule was Vostok 1, a one-person spacecraft that could also be used as a camera platform. It had an instrument panel and a seat for the cosmonaut. As a test module, it could support the cosmonaut in relative comfort and allow him or her to communicate with the ground. Further refinements for control and comfort would come later.

Paging Yuri Gagarin

The first man into space was Yuri Alekseyevich Gagarin, a senior lieutenant in the Soviet Air Force. As a young adult he learned to fly airplanes. Eventually, he was drafted into the military and sent to pilot school to fly MiG-15 fighter jets. He entered the Soviet cosmonaut programme in 1960 as part of an elite group of pilots. Their training prepared them for the rigors of launch and space flight.

Gagarin's 12 April 1961 flight showed the world that manned missions were possible, and was an even larger propaganda coup for the country than the Sputnik mission was. His mission lasted 108 minutes and comprised a single orbit around the planet. Even though Gagarin was a pilot, for this mission he was strictly a passenger. All control of the spacecraft came from automatic systems backed up by ground control. Since it was a test flight, and the first of a person into space, the mission planners had no idea whether Gagarin would be physically able to steer his craft. They therefore locked his control panel but did place an envelope with an unlock code for him to use in case of emergency.

During his short flight, Gagarin communicated with the ground by radio about what he could see, how the mission was proceeding, and what it felt like to be weightless. At the end of the orbit, the retrorockets on the capsule fired and Vostok 1 re-entered the atmosphere. When the capsule reached an altitude of 2,133 metres, Gagarin ejected from the capsule and floated to Earth on a parachute, landing some 300 kilometres off course. Despite that, the mission was a huge success, and the world was rocked by the news of this achievement. Gagarin was an instant celebrity and touted as an example of what Soviet ingenuity could accomplish.

Gagarin was the first to see Earth as a planet, and like other astronauts and cosmonauts who came after him he felt moved to report on its beauty. 'I saw for the first time the earth's shape,' he said in a speech after his flight. 'I could easily see the shores of continents, islands, great rivers, folds of the terrain, large bodies of water. The horizon is dark blue, smoothly turning to black ... the feelings which filled me I can express with one word: joy.'

Six Soviet cosmonauts flew in six missions aboard aboard Vostok. They included the first woman in space, Valentina Tereskhova. She flew on 16 June 1963. Their flights, which occurred from 1961 to 1963, proved the ability of the Soviets to put people into space, have them orbit the planet, and then return safely to Earth. Beyond these six missions, the Vostok programme had eight more flights planned, but they were cancelled and their components recycled into the Voskhod programme.

The US Responds

Challenged by the Soviet accomplishments, the US worked to launch its first astronauts to space. It was a hectic time, with spacecraft designs and manufacturing moving quickly The Mercury programme ran from 1961 through 1963, using rockets developed for military use. The first manned Mercury capsules were launched aboard Redstone rockets (based on the old V-2 design), followed by four flights atop the Atlas launcher.

Before the astronauts could fly, NASA tested the flight hardware with primates – the most famous of these was Ham the chimp.

His co-flyers were Sam (a rhesus monkey) and Enos the chimp. All three flew safely, opening the way for humans to go next.

Seven astronauts were selected for the Mercury missions. Alan Shepard flew into history as the first American to go to space on 5 May 1961. He was followed by Virgil I. 'Gus' Grissom, John H. Glenn, Malcolm Scott Carpenter, Walter M. 'Wally' Schirra, and Leroy Gordon Cooper. During these missions, NASA and the astronauts learned about safely orbiting people around the planet.

Not long after the Mercury flights, it became very clear that the Soviets would aim for the Moon. Of course, the US had announced its own aims for the lunar surface, pushed heavily by von Braun, who had read Arthur C. Clarke's 1951 book *The Exploration of Space* and found the arguments for lunar flight very intriguing. He was so entranced with the work he used it to convince President John F. Kennedy to consider a Moon shot for NASA. In 1962, Kennedy made his famous 'We choose to go to the Moon' speech and the stage was set:

> We choose to go to the moon in this decade and do the other things, not because they are easy, but because they are hard, because that goal will serve to organise and measure the best of our energies and skills, because that challenge is one that we are willing to accept, one we are unwilling to postpone, and one which we intend to win.

This bold statement occurred against a backdrop of the Cuban Missile Crisis and the deepening of the Cold War. By that time, forty-five satellites had been launched to Earth orbit, with forty of them from the US. Clearly the high frontier of Earth orbit was just a stepping stone for NASA as it, too, was pointed to the Moon by a president cognisant of both political necessity and the scientific rewards such a step would bring.

The resulting Apollo programme was planned to take astronauts to the Moon and bring them back again. Before that could happen, precursor missions would test critical flight hardware and allow astronauts to practice spacewalks and other activities. The entire enterprise required huge technological leaps and commitments of money and people.

Flying the Voskhod and Gemini Missions

After the Vostok missions ended in 1963, the Soviet Union was ready for the next steps to get their cosmonauts to the Moon. Chief designer Sergei Korolev (who was in charge of the space program) wanted dearly to get his country's cosmonauts onto the lunar surface before the US could get its people there. That required a more complex spacecraft to carry more than one cosmonaut to space, which led to the development of the Voskhod capsule. Several flights without crews were made before the first human mission left Earth on 12 October 1964. It carried Vladimir Komarov, Konstantin Feoktistov and Boris Yegorov. It was the first time more than one person had been launched into space by either country. Their mission was followed by the Voskhod 2 on 18 March 1965. It featured the first spacewalk, performed by Alexei Leonov, a feat which the Soviet Union heralded proudly, even though Leonov ran into near-fatal problems when his space suit overinflated. He was able to deflate it enough to get inside the capsule, and then he and fellow crew member Pavel Belyayev struggled to close the hatch.

The third Voskhod flight, called Kosmos 110, carried two dogs to space for a twenty-two-day mission. Subsequent human flights in the Voskhod series, which would have been a series of one- and two-person missions to study effects of humans in space, were cancelled after delays in the programme. The Voskhod programme, which was really planned to provide some flashy 'firsts' for the Soviets, was rapidly eclipsed by the successes of the US Gemini programme. In addition, the changing political climate in Moscow was not conducive to the types of 'one-off' flights the programme had provided.

While the Soviets were flying Voskhods, the US was sending Gemini capsules to space. Gemini was the training programme for the upcoming Apollo trips to the Moon. The name came from the celestial twins in the constellation of the same name. Gemini objectives were to test humans and equipment for use on long-duration (up to eight days) space flights; to practice extra-vehicular activity in space; to learn how to rendezvous and dock two spacecraft together; and to perfect the atmospheric re-entry process that would bring the space travellers back to Earth in a safe landing. Each Gemini capsule was launched aboard a Titan II rocket.

The Gemini missions carried ten two-man crews to space, including veterans from the Mercury programme, as well as a new set of astronauts selected for these and the later Apollo missions. Before NASA sent humans, however, it made two Gemini flights in 1964 and 1965 without crews in order to test the equipment.

The activities on the Gemini missions included space walks, rendezvous missions and dockings – all tasks that would be necessary for moon trips. After Gemini, NASA began the Apollo programme. However, before anyone could get to the Moon, both the US and Soviet space programs faced tragedies.

Apollo's Unfortunate Beginning

Apollo was a three-piece spacecraft, comprising a command/service module (CS/M) and a lunar lander lofted to space atop a Saturn V, which was a three-stage rocket. The first stage boosted the rocket off the ground, a second stage started up 168 seconds later and took the CS/M and lander to a parking orbit, and after several orbits the third stage sent it to the Moon. Once at their destination, two astronauts would go into the lander and head to the surface, leaving one astronaut in the CS/M orbiting the Moon. After the lunar surface mission, the two would launch and rendezvous with the orbiter, then head back to Earth.

The first unmanned test flights took place in 1966 and were called AS 201, AS 202, and AS 203. AS 204 and AS 205 were supposed to further test the hardware and command and control systems, but 205 was cancelled and 204 was reworked as the first orbital crewed test of the command/service module. Since it was going only to Earth orbit, it would ride atop a Saturn IB rocket. (Trips to the Moon required a vehicle with more power, so they rode to orbit on Saturn V rockets.) In addition to the on-orbit work, AS 204 would test all aspects of launch operations, as well as the ground-tracking stations and other control facilities. The mission was slated for a two-week run.

There were many delays in that first manned Apollo mission, which was eventually renamed Apollo 1. Planners settled on a target date of 21 February 1967. Three astronauts – Gus Grissom,

Edward H. White and Roger B. Chaffee – were chosen to fly and they worked with the spacecraft builders as part of their training.

On 27 January 1967, Grissom, White and Chaffee arrived to do a launch rehearsal. The capsule was festooned with instruments, cables and wires, with three seats tightly fitted together. It was also filled with a pure oxygen atmosphere at higher-than-normal pressure. The astronauts, dressed in full space suits, entered the cabin and took their positions. Technicians sealed the door as if the test were the real thing. Several hours into the rehearsal a power outage occurred, followed by electrical arcs and sparks. This ignited a fire which destroyed the interior of the capsule within a minute and killed the three men.

The Apollo 1 fire led to a twenty-month delay in flights while the agency fixed the problems that led to the fire. The practice of testing and flying in a pure oxygen atmosphere was something the Soviets had grappled with. On 23 March 1961, cosmonaut candidate Valentin Bondarenko was severely burned at the end of a fifteen-day test inside a chamber filled with a 50-percent-oxygen atmosphere. He had dropped an alcohol-soaked cloth on to a hotplate, which ignited. He was eventually pulled out of the chamber, but died shortly after his rescue. The information about that test was not publicly released by the Soviets due to their penchant for secrecy.

Soyuz Suffers a Setback

By 1967 the Soviet space programme had developed its Soyuz lunar missions programme, originally with unmanned test flights. Then, veteran cosmonaut Vladimir Komarov climbed aboard and launched into space on 23 April 1967. Problems set in as the spacecraft achieved orbit. Its left solar panel failed to deploy, which affected the star sensors needed for attitude control. With limited power and the spacecraft losing its stable position in space, Komarov struggled to fix the problems. He was unsuccessful, and with spacecraft systems failing, mission controllers decided to end the flight and bring him home after seventeen orbits. The re-entry burn put the capsule off course, and Komarov was instructed to manually re-orient his craft. After a partially successful attempt at another de-orbit burn, he was finally on the way home. However, problems with the parachutes that would bring it gently back to Earth turned Soyuz into a projectile dropping rapidly toward the surface. There was nothing to stop the capsule from becoming a descending bomb. It smashed into the ground, killing Komarov as the solid-fuel rockets burst into flame. This fiery disaster set the Soyuz programme back nearly a year and a half while the agency worked to fix the problems that led to the crash.

Soyuz 3 flew in 1968, carrying cosmonaut Georgy Beregovoy on eighty-one orbits over a four-day period. It was a rendezvous mission with an unmanned Soyuz 2, but the docking attempts failed. Soyuz 2 was de-orbited and Beregovoy continued the mission by working on mapping what he saw out of his window and making weather observations. He returned to Earth on 30 October 1968.

After Soyuz 3, the Soviets faced numerous technical problems with the N1 rocket, which was supposed to carry missions to the Moon. Static test fires of all the engines together never happened, and problems with the fuel valves hobbled the rocket. Full-up testing never happened due to funding problems. The N1's problems weren't confined to the technical side, either. Infighting between rocket development teams after the overthrow of Premier Khrushchev led to changes in the lunar programme plans. When Korolev died, his replacement had little political savvy and was unable to move the programme past its technical problems. The N1 problems were never solved and this led to the cancellation of the Soviet Union's manned lunar missions.

Another major catastrophe hit the Soyuz programme in 1971, at the end of the Soyuz 11 mission to the Salyut space station. The cosmonauts on this mission – Georgy Dobrovolsky, Vladislav Volkov and Viktor Patsayev – arrived on 7 June 1971 for a three-week stay. Among other things, they broadcast reports back to Earth and set a space endurance record. On 30 June 1971, the three men boarded their Soyuz capsule for the return to Earth. Unfortunately, a ventilation valve malfunctioned when the service module capsule separated from a service module; this caused the cabin to depressurise, and the cosmonauts asphyxiated just as the capsule was re-entering Earth's atmosphere.

After that mishap, the Soviets reworked the Soyuz capsules to solve the depressurisation issues and redesigned them to carry only two people. They were subsequently used on a joint Apollo–Soyuz mission in July 1975, and then carried crews to the Mir space station (built in 1986 and used until 2001) and mission. The Soyuz capsules remain in use today. In their current configuration (in 2017 and into 2018) they are only crew-carrying vehicles providing access to the International Space Station.

Apollo to the Moon

After the Apollo 1 disaster, NASA redesigned the command/service module and continued with the push to the Moon. Apollo 4 was an 'all-up' test of the Saturn V rocket, including orbital separation, the separation of the CS/M from its service module and re-entry of the command module. Apollo 5 was a flight to test the lunar module. Apollo 6 was the last unmanned mission in the series, and it was supposed to go into orbit around the Moon. Problems with the Saturn V rocket damaged the third-stage rocket, which was supposed to perform the lunar orbit injection. Instead, NASA duplicated the Apollo 4 mission profile. It worked well, and the 'okay' was given to send crews to space.

Apollo 7, carrying Donn F. Eisele, Wally Schirra and Walter Cunningham, went on an eleven-day mission around Earth to test out the CS/M with humans aboard. That was followed by Apollo 8, carrying astronauts Frank Borman, James Lovell and William Anders on an historic mission with many 'firsts'. The three men were the first humans ever to leave Earth orbit, escaping Earth's gravity. They were the first to see the whole planet from a distance, and the first to feel the tug of lunar gravity. They became the first to orbit the Moon, circling it ten times. Plus, they were the first to see the lunar far side.

Apollo 9 took astronauts James A. 'Jim' McDivitt, David Randolph Scott and Russell L 'Rusty' Schweikart to Earth orbit in 1969, where they practiced rendezvous and docking techniques between the CS/M and landing modules. That was followed by Apollo 10, which was a dress rehearsal for the moon landing. Astronauts Thomas P. Stafford, John W. Young and Eugene A. Cernan did everything that an actual lunar landing would do except actually land on the Moon. They separated the CS/M and docked it with the lunar module. They then left for the Moon, where they practiced undocking the lunar module and orbited the moon for eight hours practising the actual ascent and docking manoeuvres.

The stage was sent for Apollo 11, carrying Neil A. Armstrong, Edwin Eugene 'Buzz' Aldrin and Michael Collins to the Moon for the first human landing. On 20 July 1969, the Lunar Excursion Module (LEM) – named Eagle – with Armstrong and Aldrin aboard separated from the CS/M carrying Michael Collins and headed to the surface. They landed at Mare Tranquillitatis (Sea of Tranquillity) at 20.17.40 UTC and spent the next twenty-one hours there, stepping outside for the first time on 21 July at 02.51 UTC. The famous words 'One small step for man, one giant leap for mankind' radioed back by Neil Armstrong reached a worldwide television audience of more than 600 million people. It was a defining moment in human history.

Once on the ground, the two astronauts activated the television camera, set up the US flag, placed a seismograph and a reflector on the lunar soil, and collected more than 21 kg of Moon dust and rocks. Then, they climbed back aboard the Eagle, prepared it for the return to the Command Module and went to sleep. It was a momentous one-day exploration, and won the Space Race for the US.

The Apollo missions continued until late 1972, when Apollo 17 took the last trip to the Moon. Each of the missions (save for Apollo 13) explored a different part of the Moon, allowing scientist-astronauts to perform lunar geology experiments, photograph the terrains, and gather more samples of lunar surface materials.

Apollo 13 was an anomalous mission and very nearly a disaster. It lifted off from Earth on 11 April 1970, and was on its way to the Moon when an oxygen tank in the service module exploded, sending the spacecraft tumbling through space. The service module lost power and was unable to provide cabin heat. The crew shut down the command module and moved into the landing module, which they used as a lifeboat over the next few days. With help from ground controllers, they improvised repairs to the life-support system that removed carbon dioxide from the cabin.

The lunar landing obviously had to be scrapped, but the astronauts did loop around the Moon to get the necessary gravity assist to bring them home. As they neared Earth, they moved back into the command module and jettisoned the damaged service module and the lunar module on the way to splashdown. The crew landed back on Earth on 17 April 1970, splashing down not far from their recovery ship near American Samoa. It was a harrowing experience for the astronauts and support crew.

The end of the Apollo missions was simultaneously the end of an important step in American space efforts and the winding down of a tumultuous time in American history. The Vietnam War was nearly over, and the country faced social changes and political fallout from it. NASA moved on to new projects, almost as if the Moon were 'done' and it was time to move on. In the afterword in his book about the Apollo era, *A Man on the Moon*, historian Andy Chaikin concluded about the missions:

> Just four years after Sputnik launched the Space Age in 1957, John Kennedy challenged the nation to reach for the moon; eight years after that Armstrong and Aldrin were leaving their footprints on the Sea of Tranquillity. That was sooner than science fiction writers and space visionaries had dared to predict. Nor would they have imagined that after reaching the moon we would almost immediately pull back from it. The last lunar voyage ended just 48 months after the first began. Just like that, Apollo was over...

Decades after the landings, the Moon has not been revisited by humans. Robotic probes continue the exploration we began back then. However, renewed interest in having people return to its dusty surface is stirring among mission planners in China, the US and Russia. Still, there will never be another time quite like the first Space Race.

Soviet Aftermaths

After losing to the Americans in the race to the Moon, the Soviet Union continued to send unmanned missions such as Zond 7 and Zond 8 to survey the lunar landscape and Luna 16 and 17 to bring back rock samples and travel the surface with a robotic rover, respectively. Their robotic lunar explorations continued through the 1970s.

The country's space priorities shifted radically. For the Soviet leadership, sending cosmonauts to the Moon had lost propaganda value in the wake of the Apollo triumphs. Changes in the Soviet economy and both political and scientific leadership made it difficult to continue trying to send their people to the Moon. Comments made by government spokesmen in the late 1960s, particularly after the landing in 1969, seemed to suggest the Soviet leadership wanted to save face by claiming there never had been a race to get people to the Moon. It had to be a disappointing end for the Soviets, but to their credit they simply pivoted and faced in another direction, plunging into three decades of largely successful space station developments that lasted through the collapse of the Soviet Union and the rise of the Russian Federation in 1991.

Planetary Exploration during the Space Race

The Apollo and Soyuz missions were not the only space efforts taking place in the years of the Space Race. While they taught valuable lessons for sending people to low Earth orbit and to a nearby world, distant destinations beckoned. The same technology that sent rockets to space and men to the Moon was used to deploy robotic probes to other worlds of the solar system. Venus, Mars and Mercury missions were built and launched. The Soviets concentrated largely on Venus and Mars in the years after the lunar missions, while the US planned missions to nearly every planet. In the decades of the 1980s and 1990s, every planet but Pluto was explored (its flyby occurred in 2015), and missions to comets and asteroids were on the drawing boards for the 1990s and beyond.

The rationale for planetary missions is a fairly simple one: to understand more about how they (and Earth) formed, it's necessary to observe them *in situ* – to characterise their landscapes and atmospheres, magnetic fields, gravitational pull, moons (if any), presence (or absence) of water, and more. The search for life, or life-friendly habitats, underlies much of planetary exploration. Certainly there was (and remains) a great deal of interest in finding places where humans could live beyond Earth. The Moon is a difficult environment (without air, though it does have water ice at its poles). Mars seems more earth-like in certain ways, and remains an important focus of research for future human missions.

However, the search for life native to other worlds is equally important. Could living beings exist beneath the icy surfaces of

Jupiter's moon Europa, or Saturn's moon Enceladus? Did life once thrive on ancient Mars? For that matter, how did life start on Earth? These questions don't stem just from biological curiosity; they are central to the study of geology and chemistry as well. Understanding how other worlds form and perhaps spawn life may tell us much about our own world; conversely, what we know about our planet helps us understand more about how other planets and moons have formed and evolved.

Not all of our solar system probes have been aimed at other planets; some also observe Earth's atmosphere, weather and landscapes. Their data allow scientists to make better models to help understand changes to our climate and learn more about how they affect landforms, crops, animals and human life. Planetary science is an important part of space exploration, and we will look at some missions in greater detail in chapter 7. (See Appendix D for a timeline of planetary missions.)

Space-based Astronomy

The rise of space-based astronomy is another major accomplishment of the Space Age. The reason for observing the cosmos from space is a simple one: it's a better view from above our atmosphere. Astronomers have long known that our planet's blanket of air interferes with light from distant objects as it passes through. In some cases, it's enough to make the view through a telescope very blurry. In other cases, the blanket of air actually blocks or absorbs ultraviolet, much of the infrared band, x-rays and gamma radiation. The result is an incomplete view of the universe, almost as if we were looking out to space wearing blinkers.

Scientists used high-altitude sounding rockets and balloons for short surveys of the sky starting in the late 1940s, but what they really wanted were orbiting observatories to perform long-term surveys and observations. Rockets provided access to orbital space. Satellites launched in the early 1960s focused on such objects as the Sun, followed by satellites with detectors sensitive to gamma rays, x-rays and ultraviolet light. Today, space and solar telescopes regularly make observations of celestial objects out at the limits of detectable space.

Nearly a hundred space telescopes and particle detectors have been launched since the late 1960s. The Orbiting Astronomical

Observatory (OAO-3) observed ultraviolet light targets in the universe even as it was watching the Sun. Over the years, observatories from the US, China, various European countries, India, Japan, Korea and the Soviet Union/Russia have explored the universe from Earth orbit and near-Earth space. Among the more famous are the Cosmic Background Explorer (COBE), the Herschel Space Observatory, the Hubble Space Telescope (HST), the Kepler telescope, the Spitzer Space Telescope and the Chinese Insight Hard X-ray Telescope.

Multinational teams of scientists work together to plan and implement these missions, and then they study the data returned. International agreements and projects spread the costs among all participants. We will look more closely at these astronomy and astrophysics missions and their discoveries in chapter 8.

Space Stations and Shuttles

The Space Race has produced more missions than the trips to the Moon and probes to the planets, and produced other fascinating possibilities in the form of space stations and aerospace planes. The idea of putting humans in permanent orbit around Earth, while not a new one (it has long been a staple of science fiction stories), was beginning to look more and more realistic.

The Soviets wanted to get ahead of NASA and began working on the Salyut programme. It began as a tightly guarded military programme called Almaz, created to build secret space stations for use in reconnaissance. The programme extended through 1982, across a half dozen space stations using the name Salyut. All were occupied and used except for Salyut 2, which was wracked by an explosion shortly after achieving orbit. Salyuts 3 and 5 were largely dedicated to military operations, while 4, 6 and 7 were more science-oriented. International crews were invited to work aboard Salyut 6 for short periods of time.

Salyut 1 was launched on 19 April 1971, and the first crew arrived at the station aboard Soyuz 10 but were unable to dock with the station. They ended the mission and returned to Earth. The first successful visit to Salyut 1 came when the crew of Soyuz 11, Georgy Dobrolsky, Vladislav Volkov and Viktor Patsaev, arrived on 7 June of that year. They stayed for twenty-two days, setting the first endurance record for living and working in space, and transmitted daily reports back to Earth.

Unfortunately, as mentioned earlier, that mission ended in tragedy. On 30 June, the crew climbed back into their Soyuz craft and left the station for Earth. Just at atmospheric re-entry, the cabin lost pressure at an attitude of 167 kilometres. The three men, who were not wearing pressure suits because there wasn't enough room in the cabin, died within just a few minutes. The accident led to a lengthy investigation and redesign of the spacecraft so that only two cosmonauts could ride in it, wearing space suits in case another mishap.

Mir

The ongoing redesigns and refinements of the Salyut stations led to the development of the Mir space station, which was deployed in low Earth orbit in 1986. Its name means 'Peace' and it was (for its time) the largest artificial satellite orbiting our planet. It hosted successive crews of astronauts, scientists and others during its ten-year mission, and welcomed eleven US space shuttle dockings. Mir was launched by the Soviet Union and then taken over after the country's collapse by the Russian Federal Space Agency.

As the station grew older, it suffered equipment failures, systems problems and accidents. Its original five-year mission had stretched to fifteen when the Russians made the decision to de-orbit, in part because they were committed to supporting the International Space Station. On 23 March 2001, the station entered Earth's atmosphere and burned up, with pieces falling into the South Pacific.

Skylab

While the Soviets were building and deploying their Salyut stations, NASA launched its first space station, called Skylab. It was the replacement for a cancelled military mission called the Manned Orbiting Lab. NASA announced it before the end of the Apollo programme as a way to continue the US dominance in space. Committees within NASA identified both scientific and national security rationales for an orbiting station. First, it would help the agency develop technologies for the next generation of space vehicles. In addition, technologies of the Apollo era could be repurposed to the space station, and long-term space habitation would require the development for advanced medical and life-support technology. The science case for Skylab included astronomy and Earth observations. Of course, the Soviet presence in orbit was another obvious driver for Skylab.

Skylab went into orbit on 14 May 1973, launched by a Saturn V rocket. Despite major damage to its sunshields and solar panels, the station went on to host three manned missions. During the years it was occupied, the crews performed space walks (which helped establish guidelines for similar walks during the construction of the International Space Station), did more than 2,000 hours of scientific experiments, and studied the effects of microgravity on the human body.

There were plans to extend the life of the station and boost it to a higher orbit, but the mission to do that, Skylab 5, was cancelled. As NASA held internal debates about what to do with the station, one idea floated was to have space shuttle crews visit for repairs and reactivate the station. Due to the cost of such missions, they never got beyond the drawing board. The station was de-orbited on 11 July 1979.

The International Space Station

After years of Soviet space stations, plus the American Skylab, five space agencies worked together to build the International Space Station (ISS). NASA, Roscosmos (the Russian space agency), JAXA (Japan), the European Space Agency (ESA) and the Canadian Space Agency (CSA) worked together to conceptualise and build it.

The ISS is the largest space construction project yet undertaken. Its assembly began in November 1998 with the launch of the Russian Zarya module. That was followed by the US Unity module, brought by NASA's space shuttle. In all, the station has fifteen pressurised modules contributed by the US, Russia, Japan, the ESA and Italy. Construction was complete in 2011.

Today, the ISS is a regular sight in Earth's skies. At any given time at least four people are aboard, performing science experiments. The station is continually serviced by autonomous supply launches, and crew members have been flown to and from the station by Russian Soyuz craft and shuttles.

Tiangong

The ISS is not the only station on orbit. The Chinese Tiangong 1 orbiting lab was launched in 2011, and was visited by crews in both 2012 and 2013 before being abandoned. Its follow-up, Tiangong 2, was launched in 2016, performing as a test bed for technology to be used on a future permanent space station. The next Chinese space station modules (called Tianhe-1, 2 and 3), each weighing 20 tons, will be lofted to space on the Long March 6 rockets. The complete station will require perhaps a dozen launches. Eventually, the station and its scientific test modules may be open to crews from around the world.

The Shuttle Age

The dream of flying a 'space plane' to orbit was long a staple in science fiction stories and movies. One of the most graceful concepts illustrated the poster art for the movie *2001: A Space Odyssey*, with a winged space plane on approach to an orbiting space station. For many, it seemed like the ultimate science fiction idea come true when both NASA and the Soviets announced plans to build winged shuttles. The US shuttle programme was long on the drawing boards. It began formally in 1972, with the idea of making it a sort of 'space truck' to bring crews and cargo to low Earth orbit.

The Soviet shuttle, called *Buran*, was planned as a reusable space system by the Central Aerohydrodynamic Institute, a Soviet think tank for aircraft design. The ideas for *Buran* were first developed there, and several high-altitude jet aircraft served as test beds for ideas later implemented in the Soviet shuttle. When the US announced its shuttle programme, the Soviet development kicked into high gear. This included the study of plans for the American shuttle, which was an unclassified project. The Soviets automatically assumed the US shuttles would be carrying laser-deployed weapons to space. This influenced their shuttle design to a large degree. The assumption made sense, with then-President Reagan's embrace of so-called 'Star Wars' technology.

The Soviet agency began building a fleet of full-scale *Buran* orbiters in 1980. Only one ever flew to space, on an unmanned mission on 15 November 1988, lofted to orbit by an Energiya rocket. With the collapse of the Soviet Union, the *Buran* programme was placed on indefinite hold and eventually cancelled altogether. The orbiter that flew, OK-1K1, went into storage and was never used again. In 2002, the hangar where it was stored collapsed and destroyed the spacecraft. The others were abandoned in hangars.

NASA built five shuttles that were used for 135 missions between 1981 and 2011. They were *Atlantis, Columbia, Challenger, Discovery* and *Endeavour*. The test orbiter *Enterprise* never flew in space, but was used in various drop tests and landings, and supplied parts to other orbiters as needed. *Enterprise, Atlantis, Discovery* and *Endeavour* are now museum exhibits at the Intrepid Sea, Air & Space Museum in New York, the Kennedy Space Center (Florida), the Smithsonian's Steven F. Udvar-Hazy Museum near Washington, D.C., and the California Science Center in Los Angeles, respectively.

The shuttles carried crews to orbit and then returned them to Earth for 'rolling landings'. Each one made multiple flights, lofted by two solid-rocket boosters (SRBs) and orbiter engines. They carried between two and eight crew members, plus pieces of the International Space Station, science experiments, telescopes, labs and satellites. Among the highlights of the shuttle era were multiple dockings with the Soviet Mir space station, deployment of the Hubble Space Telescope (plus its servicing missions), the Chandra X-ray Observatory, the Magellan spacecraft to Venus, Galileo to Jupiter, and Ulysses to orbit the Sun. Shuttles also brought Spacelab to orbit, where scientists performed medical, astronomical, physics and materials-science experiments during short missions. In all, 180 satellites and other payloads were carried to space aboard the shuttles.

It's easy to see how successful NASA's $196 billion shuttle programme was just based on the missions it performed. It carried 355 individual astronauts from sixteen countries to space. Shuttles docked at Mir eleven times and thirty-seven times at the ISS during and after its construction. They were the US's main ride to space for astronauts and cosmonauts, and performed for three decades.

Tragedy struck two shuttle missions. On 28 January 1986, space shuttle *Challenger* (STS-51L) was destroyed seventy-three seconds after launch when components of the solid rocket boosters failed, causing an explosion that tore apart the orbiter. All seven crew members perished in the accident, which led to a more than two-year moratorium on shuttle flights and delays in major programs, including the deployment of the Hubble Space Telescope. The space shuttle *Endeavour* was built to replace *Challenger*.

Space shuttle *Columbia* (the first orbiter to fly, in 1981) was destroyed on re-entry over Texas on 1 February 2003 during mission STS-107. It broke up after hot gases penetrated a wing of the orbiter, which had been damaged by a collision with a piece of foam during launch. The orbiter lost its control systems and disintegrated. All seven crew members were lost. Flights were halted for two years as NASA investigated the second tragedy. During that time, the International Space Station was serviced by Russian spacecraft, and the station was manned with a crew of only two.

The space shuttle programme was originally slated to last for fifteen years, but the requirements of the ISS construction and other missions (such as servicing HST) kept the NASA fleet flying for thirty years. However, they were aging, and in 2004 then-President George W. Bush directed that the shuttles be retired. A new programme, the Vision for Space Exploration, called for a more capsule-like approach, as well as ideas for new winged aircraft. To many, the day of the 'space plane' was thought to be over. The final launch in the programme, by *Atlantis*, took place on 8 July 2011. After its return, the orbiter fleet was decommissioned.

On the military side, the X-37B orbiter (which looks like a miniature space shuttle), built by the Boeing Corporation for the US Air Force, is a long-running reusable space plane. It flies robotic missions for reconnaissance and other purposes. The X-37B was originally a NASA project that was transferred to the US Department of Defence in 2004. Two of these orbiters have flown four missions, for a combined time of well over 2,000 days in space, launched on Delta and Falcon 9 and making autonomous landings. The missions are largely classified, but are known to include equipment and materials testing for rocket motors, communications technology for upcoming civilian missions, and materials for research on the effects of long-duration flight on the orbiters themselves.

The Global Space Age

While the two main players battled it out during the Apollo era, other countries were busily creating and expanding their own spacefaring efforts. With the realisation that maintaining separate national space efforts was an expensive proposition, leaders looked toward cooperative agreements in scientific research in space.

Great Britain, for example, began developing a series of satellites, working with the NASA Goddard Space Flight Center to build the first British satellite, called Ariel 1. It was launched from Cape Canaveral on 26 April 1962 to study solar radiation and Earth's topmost atmospheric layer. France launched its first sounding rocket in 1952. The first of France's space-bound rockets, the Diamant, launched in 1962. Japan, a latecomer to the space scene due to restrictions placed on it after the Second World War, participated in the IGY with its own first suborbital rocket launch.

Today, more than seventy governmental space agencies have been formed around the world, including the European Space Agency, which unites twenty-two member and advisory states in a common space effort. Thirteen agencies around the world have the ability to launch rockets and payloads to space, and as of 2017, Russia and China are the only ones capable of launching humans to space.

In addition to scientific work, countries have always been interested in controlling their own access to satellites. This has led to tremendous growth in the global communications and reconnaissance satellite industries (more on this in chapter 6).

The Space Age Continues

As of this writing, in late 2017, countries around the globe continue to plan for near-future space exploration. Space is no longer an exclusive domain, although it remains an expensive one to explore. Savvy political and business leaders see it as the next step in technological, social and political development – even as some countries still struggle with economic progress. Nonetheless, the heady days of the Space Race have given way to a long-term Space Age. There are concentrated efforts to make space access more affordable, frequent and available to anyone who needs to get something to low Earth orbit or beyond. The leap to space has changed many of the world's economies and cultures forever. It has expanded our view of the universe in ways few could have expected back when rockets were developed and missions were still on the drawing boards.

3

Human Steps to Space

We will certainly see teachers, journalists, artists and poets in
space. Whatever it takes to be the best is what it will take to get
you into space.

Astronaut Eugene Cernan

Space is for everybody. It's not just for a few people in science
or math, or for a select group of astronauts. That's our new
frontier out there, and it's everybody's business to
know about space.

Christa McAuliffe, teacher and Challenger astronaut

I had the ambition to not only go farther than man had gone
before, but to go as far as it was possible to go.

Captain Cook

Imagine climbing aboard a spaceship headed to orbit. Your mission:
head to Mars for extended exploration. You stow your gear and get
strapped into your seat. Your instrument panel is in front of you,
plus a video display showing the outside world. You put on your
headset and go through the pre-flight check with the team leader,
whose voice guides you through the instrument and systems checks.
You've trained for over a long time, and you're ready to go.

Before you know it, you've cleared the launch pad and are
moving at more than 161 kilometres per hour straight up into
the sky. You're pressed against your seat fighting to breathe as

G forces make you up to three times heavier than your earthbound weight.

On the headphones, you hear the chatter of the people at launch control and the voice of the ship's AI giving system information.

Launch plus 20 ... all systems nominal ...
Engage second-stage rocket ...

Eight minutes into the flight, the rockets shut down and suddenly you're in orbit ... and weightless. Your head feels funny, and your body feels like it's almost not there. Fellow team members look at each other and give the thumbs-up as the mission director keeps talking.

Mars One, prepare for trans-Mars injection sequence in 30 minutes.

The mission is 'go' and Mars is your next destination.

This may seem like a scene out of science fiction, but that hasn't stopped thousands of people from experiencing it at Walt Disney World's Epcot theme park, in a ride called *Mission: Space*. It comes fairly close to simulating much of what astronauts have really experienced at launch pads at the nearby Kennedy Space Center and at Baikonur Cosmodrome in Russia. For most people, it's the closest that they will get to an actual space flight, until regular tourist trips to low Earth orbit or the Moon become a reality. However, throughout history, dreams and fantasies have become reality.

There's a good reason why so many science fiction stories focus on people taking trips to the Moon and beyond: space travel is one of the grandest adventures anybody can imagine. Leaving Earth, visiting other worlds. Yet, for all of the visions of space exploration we enjoy in *Star Trek*, *Star Wars*, *Battlestar Galactica* and *Dr Who*, and in books and pulp magazines, the actual experience of space flight – as experienced by the men and women who have done it since the 1950s – isn't always like the SF visions.

For NASA and the Soviet space programme, the technology to deliver people safely to space (and bring them back again) had to be built from existing aeronautical designs, or created from scratch, developed on the fly.

In much of early science fiction, men did the flying and piloting – women were rather less obvious in many stories. The same was largely true in the early days of the Space Race, with one major difference: in the Soviet Union, women pilots were recruited early on, and the Soviets were the first to send a woman to space, very early in their programme. In the United States, highly qualified women were recruited but discarded. It took decades for NASA to send a woman to space, and the Russians didn't send any more after the first two until 1994. By then, female commanders and ship captains were commonplace in stories, TV shows and movies.

Many tales of the future simply assume space travel is a pretty normal and common thing. In Robert A. Heinlein's *Future History* series, for example, rocket-based travel is *de rigueur*, transporting people and goods to and from Earth and other solar system destinations. He even has a spaceship pilot commandeering an entire interstellar ship in order to save a population of human beings threatened with extinction on Earth.

The contrast with our SF flight dreams and reality is stark. Today only a handful of people have set foot on the Moon, and the only long-term occupation of space has been aboard space stations in low Earth orbit. There aren't yet human colonies on Mars, nor are there research bases on the Moon. Yet we continue to look outward, lured by the chance to explore new places and the possibility of finding life elsewhere in our solar system and beyond.

The search for life is very much a work in progress. Today, we are still devising the methods of determining if, how and when life existed on other worlds of our solar system, and whether we can detect clues to life on planets around other stars. Such discoveries will surely change our outlook, much as the first space flights of the 1960s did.

Arthur C. Clarke published many non-SF stories and books relating to space travel, an idea which he helped popularise. As mentioned elsewhere, he predicted one of our best-known space-related technologies, the communications satellite. His futurism predated real events, but also took inspiration from others. Clarke, Heinlein and many others helped set the stage for public acceptance and interest in space flight, even though their stories foresaw a future far beyond what anyone was capable of achieving in the 1960s, when humans began going to space. Today, we more or less routinely send people to live on the International Space Station for months at a time, but we still lack the ability to go to the stars. In that sense, our science fiction stories still far outstrip our abilities. Missions to Mars and beyond will be grounded in the reality of what's been learned by the hundreds of astronauts who have trained and flown to space. Heading out to Alpha Centauri remains a distant dream.

At some point people will be making trips to the planets and beyond. Who will they be? What training will they need? As it turns out, the people doing the flying aren't always the heroic and dramatic figures portrayed in our literature and media. They are the first to remind us that they're simply ordinary people with incredible training and the will to explore. And yet, they do extraordinary things most people can only dream of doing.

Becoming an Astronaut

Astronauts and cosmonauts are among the most admired people in the world. The men and women of the world's astronaut corps regularly train for years to fly and work in space and know the risks and rewards. For astronauts, climbing aboard a spacecraft and being flung off Earth to space represents an exciting and challenging career. Ever since the first days of the Space Age, kids have grown up wanting to travel in space as these people do.

The first astronauts were fighter pilots and test pilots, men who knew how to fly high-altitude jets. They were military officers, for the most part, and they knew going in that they'd be test piloting spacecraft going beyond the bounds of Earth. Such people were used to a chain of command, as necessary in space flight as it is for military units working together on land or at sea.

In the beginning, both the US and Soviet Union looked toward their respective air forces to supply likely astronaut candidates for the early flights. China still selects its astronauts from the ranks of the military. (The Chinese term for astronaut is 'Yǔháng yuán', and they can also be referred to in Mandarin as 'taikonauts', or 'Chinese astronauts'.)

Later on, as Skylab, Salyut, Mir and the ISS were built and deployed, the military training requirement was dropped for specialists flying on science missions. Today, not everyone who flies to space is a member of the military, but they are highly trained in the specifics of their missions, the craft they fly in and other aspects of their jobs. The only exceptions are those who have paid to fly on short-term missions (such as Dennis Tito, the first space tourist, and Anousheh Ansari, the first female private explorer), or who pulled political strings, such as the former US senator Jacob 'Jake' Garn.

Each country's space agency sets specific standards for its astronauts. At NASA, for example, commanders and pilots must be highly flight-qualified, and have at least a bachelor's degree in engineering, biology, maths, physical sciences, or some combination of these. Chinese astronaut candidates are chosen from the country's air force and are required to have a university degree and extensive flight experience. Russian pilots who become cosmonauts usually have extensive military experience, often coupled with scientific or engineering backgrounds.

At NASA, scientist astronauts must have rigorous academic training and experience in their chosen fields. Payload specialists are those people who have very specific experience with instruments, telescopes or lab work necessary for a given mission.

Regardless of background or country, astronauts must be in top physical shape, and able to withstand both the rigor of lift-off and the weightlessness of space. In all cases, whether the astronaut candidate is a pilot, mission or payload specialist, they must meet physical requirements, including an established height range (between 157 and 190 cm for pilots and commanders, 147 and 193 cm for others) and good visual acuity and healthy blood pressure. There are no age limits for NASA astronaut candidates, although most are between the ages of twenty-six and forty-six. Other countries have similar physical requirements. In China, candidates must be between the ages of twenty-five and thirty, between 160 and 172 centimetres tall and weigh between 50 and 75 kilograms.

Beyond the physical and educational requirements, people who go to space usually fit a rather specific psychological profile. They don't usually have the dramatic and cocky 'Captain Kirk' style of spacecraft and crew management, although having a healthy sense of self-confidence *is* important. They need to be able to cope with physical stresses, sensory deprivation, disorientation and other problems of space flight. They also need to have the mental capacity to face danger and isolation while in space.

Because space flight is largely a cooperative enterprise during missions, space agencies look for people who can work well in teams, have significant problem-solving skills, can be trainable, self-reliant when necessary, and handle significant levels of stress. These characteristics also describe military, law enforcement, and emergency services personnel, and can also apply to scientists. On top of all that, astronauts need to be able to function in the public eye, since they represent their countries and agencies. This was quite important in the early days of the Space Race, when both the US and Soviet Union proudly showed off their astronauts before and after each mission.

The Environment of Space

Long-duration missions require a more nuanced understanding of how people function in space. Long-duration fliers Valeri Polyakov (438 days), Mark Kelly (522 days), Peggy Whitson (534 days) and Gennady Padalka (nearly 900 days) are helping mission planners understand the long-term effects of microgravity on the human body. When humans embark on missions to Mars and beyond, the experience of astronauts who have spent months and years in space will be invaluable in terms of understanding the physical, mental and psychological stresses they experience.

Living in space is very different from living on Earth, where we have air to breathe in an atmosphere that shields us from much of the harmful radiation coming from the Sun and outer space. Our bodies evolved in Earth's gravity, which had an influence over our musculoskeletal structure, as well as our cardiorespiratory systems.

Earth's environment is very hospitable to us and a wide variety of life, while space is a veritable desert. There is no air in space; it's a vacuum. Temperatures range from very low to extremely high, well beyond our ability to withstand. Habitats must be built to shelter the fragile lives of the astronauts. To work 'outdoors', in space, astronauts must wear special suits that supply air, water, heating, cooling and protection from radiation.

Radiation hazards in space are much greater than on Earth. High-energy particles from the Sun, as well as cosmic rays from beyond the solar system are the chief sources of radiation in near-Earth and interplanetary space. On Earth, our magnetic field shields us from the worst of the ionizing radiation. In low Earth orbit, astronauts are also protected by the magnetic field, although they do receive a slightly higher amount of cosmic rays. Particularly strong blasts of high-speed particles from the Sun during flares or coronal mass ejections have forced ISS inhabitants to seek shelter in shielded sections of the station for short periods of time. Astronauts traveling to Mars or living on the Moon will not have Earth's magnetic field for protection and will face increased radiation exposure. They could be faced with increased risk for developing cancer.

Finally, there's the issue of gravity. The farther away from Earth you travel, the less of a gravitational pull it exerts. To live in space is to live in microgravity, with effects on the body that mission planners and medical personnel are still studying. While spacecraft and orbiting stations can carry supplies of food, water and air to make the environment as Earth-like as possible, they can't supply 1G of gravity. That, so far, is still beyond our ability to provide on an orbiting space station. Some future designs for Mars-bound missions have suggested spinning the spacecraft to create artificial gravity, which would help the travellers acclimatise to the gravity of their destination as well as maintain better health during the flight.

Training for the Rigors of Space Travel
In his 1984 book *Entering Space*, former astronaut Joe Engle described the launch experience aboard a space shuttle:

> For the crew, the journey begins with a rush of noise and a surge of motion. An instant after the tower disappears from view, the astronauts feel the launch vehicle roll into a high arc over the Atlantic ... the vehicle accelerates to a force of 3 Gs (three times the force of Earth's gravity) within seconds after launch, and it quickly becomes impossible to distinguish any direction but forward. As the rockets lunge, the crew members are pushed back hard against their seats; their bodies feel very heavy, and they can move their arms only with deliberation. The vehicle vibrates wildly, and in spite of the crew's protective helmets, the cabin

is incredibly noisy. Within a span of 30 seconds, the blue sky
outside the orbiter's windows has become a deep and solid black.

That is the transition faced by every person who goes to space: the
ordeal of launch. Leaving our planet's gravity well is a tough ride.
Beyond that, living and working in space is, without a doubt, a
dangerous and risky business. It requires a mentally and physically
fit astronaut corps whose members are in superb condition. During
launch they must to be able to function wearing a space suit and,
if necessary, take command of a spacecraft. Once they get to space,
they rely on their discipline and training to immediately get to work
on the tasks they need to fulfil their missions.

Physical training is a large part of the regimen of astronaut life
and it's not uncommon to see astronauts running and swimming as
part of their exercise routine. As important as it is to be in shape
for launch and deployment, astronauts also train physically for the
long term. They need to be in good shape because space travel has
adverse effects on their bodies. Once in space, particularly for more
than a few days, they are required to exercise each day to keep their
heart, lungs, muscles and bones healthy.

The effects of space travel on the human body are still being
understood. Humans evolved in a 1G environment; going to
space forces one's body to adapt to no-G or low-G very quickly.
Muscles begin to atrophy after just a short time in microgravity.
Current and past astronauts have all reported on how their bodies
changed during their missions. Bones lose mass (a condition called
'space flight osteopenia'); about 1 per cent for every month spent
in space. Other issues include changes in cardiovascular function,
fluid redistribution in the body, sleep and immune system disorders
and changes in eyesight.

Studies of changes in astronauts' eyes in particular show that
the eye is reshaped by trapped internal fluids, causing a condition
called intracranial hypertension. The condition appears to affect
men more than women, and NASA is working on understanding
how long it persists after flight. Nonetheless, it appears that a very
long stay in space – whether a year aboard the space station or a
year out to Mars – poses specific threats to a person's health. How
long those effects last and whether they can be mitigated during

or after flight is something that space agencies are continuing to study. It is up to future mission planners to figure out ways to keep astronauts healthy on their way to Mars and beyond.

Any astronaut candidate (called an AsCan in the US) goes through training drills designed to acquaint them with the sensations and experience of launch and the transition to a microgravity environment. This includes riding in high-G centrifuges, undergoing underwater training and experiencing simulated weightlessness in parabolic-arc flights aboard NASA's KC-135 (the 'Vomit Comet') jet aircraft. Similar 'basic' training also occurs for astronauts working at (or with) space agencies in China, Europe, India, Japan and Russia.

Learning to be a Spacefarer

Today's astronauts and cosmonauts are required to have a college education and several years of experience in their field of specialty before they can even apply for a slot at their country's space agency. As mentioned above, those academic qualifications are mostly in the sciences or mathematics.

After basic training, an astronaut candidate at NASA, for example, undergoes two years of training specific to their missions and assignments. In Russia, the second level is 'group training'. The third level in both countries is more mission-specific and is called 'increment-specific' in Europe, 'crew training' in Russia, and 'intensive training' at NASA.

Astronauts learn to fly aircraft (if they don't already know how), and they also train on mock-ups of the ISS and Soyuz launch hardware. Would-be cosmonauts and spacefarers from other countries also undergo similar training, as do astronauts from Canada, Europe and other countries from around the world. For Soyuz launches astronauts also learn the Russian language, and some crew members undergo medical training or specialise in life sciences as well.

Astronaut training includes a lot of classroom time as well as simulator training. For example, the five maintenance and repair missions to the Hubble Space Telescope, undertaken from 2003 to 2009, required hundreds of hours of astronaut preparation. Not only did they need to study the telescope's schematics and deployment, they also had to understand how the spacecraft worked as a unit and what each of its instruments did.

Each Hubble Space Telescope (HST) servicing mission had very specific tasks to carry out, and every one was rehearsed over and over again. The neutral buoyancy tanks at the Johnson Space Flight Center in Houston, Texas were outfitted with a full-scale mock-up of Hubble's main body – its 'bus' – and astronauts used it to simulate tasks in relative weightlessness. There, they learned to remove massive and unwieldy cameras and replace them with newer instruments. Among other things they practised replacing gyros, fixing electronics panels and putting on new solar panels. The repetition of the repair projects gave each astronaut an almost physical 'memory' of the tasks to be done long before they ever got to orbit and had to bring the telescope to the shuttle bay.

Living and working in space is physically challenging. Motions taken for granted on Earth take on new meaning in space, where the slightest tug or push on an object can send it (or the astronaut) careening through the microgravity environment. Safety is a priority, and so much of the training focuses on learning to move in space, to respond quickly to dangerous situations (fires on the station, for example), and other emergencies. Astronaut John Grunsfeld, for example, has often described working inside the HST while on orbit, and his helmeted head was often within just a few inches of the main mirror. Years of training taught him exactly where he could put his head and how far he could move within that confined space to avoid damaging the telescope. Learning to work in space is made up of countless moments like John's. It's a necessary part of living and working in an environment for which our bodies were not designed.

In recent years, astronaut training centres at NASA and other agencies have adopted immersive, virtual- and 'hybrid'-reality technology to help prepare astronauts for the missions they will fly. The VR training team maintains an up-to-date model of the ISS, for example, where astronauts can learn about the layout of the equipment. Extravehicular activity training teaches astronauts how to do space walks, and other systems give them experience in using robotic instruments such as the space station arms and robotic manipulator systems. Because accidents happen, the VR simulations also offer training on the Simplified Aid for EVA Rescue (SAFER). This is a small propulsive backpack system used during spacewalks to provide a way to return to the station if an astronaut gets stranded while outside.

Some of the VR training is done with head-mounted displays, while other simulated environments require extended movement throughout an exercise. In those cases, CAVE (Cave Automatic Virtual Environment) systems display visual cues on video walls. As noted above, a great deal of 'weightless' training takes place in neutral buoyancy tanks outfitted with mock-ups of whatever systems the astronauts are being trained to use.

One of the best-known weightless environments is the aforementioned 'Vomit Comet', a series of specially outfitted aircraft that fly astronaut trainees in parabolic arcs. Participants ride up to the top of a parabolic arc; then the aircraft dives, and its occupants experience free fall for a short period of time. It simulates the zero-G conditions astronauts work in, and helps them learn to move their bodies and manipulate tools and massive objects. For many, it's also a way to learn how to deal with nausea and other discomforts that come from zero-G. Canada, the European Space Agency and Russia operate such training flights for their astronauts. In addition, private companies in Europe and the US offer similar flights for both training and entertainment purposes. The best known is the Zero Gravity Corporation. In the US it supplies the service to NASA.

Although most space training has taken place through the space agencies in their respective countries, the realities of commercial space travel and the needs of military and civilian pilots have created new training opportunities. In 2007, the National Aerospace Training and Research Center (NASTAR) opened in Philadelphia, PA. It offers space training, research and educational tours, and partners with NASA, the FAA and other government agencies, as well as universities, laboratories and military personnel from around the world. Other training facilities include Waypoint 2, which is aimed at readying commercial astronauts for private aerospace ventures. The space tourism industry may open up further need for training the personnel who will fly tourists to space for short vacations.

Meet the Spacefarers

More than 560 people have trained as working astronauts, and most have been to space. There are more than 100 astronauts and astronaut candidates working at NASA (as of late 2017), plus a group of nineteen from other countries working with the agency. Russia has had sixty-eight cosmonauts in its corps over the years. Other astronauts have come from Afghanistan, Australia, Belgium, Britain, Bulgaria, Brazil, Canada, Cuba, Czechoslovakia, Denmark, France, Germany, Iceland, India, Israel, Italy, Kazakhstan, Malaysia, Mexico, Mongolia, Netherlands, Peru, Poland, Romania, Saudi Arabia, South Korea, Spain, Switzerland, Syria, Ukraine and the United Kingdom. China has trained nearly forty astronauts, and as of mid-2017 ten have flown missions for the Chinese space agency.

Not every astronaut dreamed of becoming one as they were growing up, but nearly all of them caught the bug when they began to study science or aviation. Every person who has flown has made incredible accomplishments, ranging from being the 'first' to do something in space to leading the development and deployment of space installations. But, like many other gifted and talented people, their interests and passions often go beyond space. Some are accomplished artists; others excel as athletes. More than a few are authors, while some have gone on to work in business and education after their astronaut stints. Many take time from their lives to talk to school children about their careers. From Alan Bean's fantastic space art depicting astronauts on the Moon and Alexei Leonov's paintings of Earth's atmosphere, to Chris Hadfield's musical talents demonstrated aboard the ISS during his stint in space and the many fine books by Buzz Aldrin, Clayton Anderson, Scott Carpenter, Michael Collins, Mike Mullane, Sally Ride, Alan Shepard, and many others, these space travellers represent a wide cross-section of talents and abilities. It's worth taking a closer look at some of the 'standouts' among the astronaut corps of the world to get to know the kind of person who lives and works in space.

Yuri A. Gagarin: First Man to Space

Each year, space enthusiasts celebrate 'Yuri's Night' on 12 April, in honour of the first man to travel into space. Gagarin was a pilot and member of the Soviet Air Force. He entered cosmonaut training in 1960, along with nineteen other pilots. Eventually he was chosen as one of the 'Vanguard Six', the cosmonauts who would go on to train for the upcoming Vostok programme. His 1961 flight caught the world by surprise and made him an instant celebrity, but, as with many astronauts and cosmonauts, he appeared to be fairly low-key about his accomplishment. In an interview after his return he praised all the people who made the flight possible, boasting just a little bit about the future by saying, 'We plan to fly some more and intend to conquer cosmic space as it should be done. Personally, I would like to fly some more into space. I like flying. My biggest wish is to fly toward Venus, toward Mars, which is really flying.'

Unfortunately, Gagarin never went to space again. After some years traveling the world as the poster boy for Soviet space accomplishments, he served the government for a while before returning to Star City to work on design and development of future spacecraft. Eventually he took over as commander of the cosmonaut training centre, and started flying fighter jets while working on an advanced degree in aerospace engineering. In 1968, Yuri Gagarin was on a routine training flight in a MiG-15UTI fighter jet with his flight instructor Vladimir Seryogin. When they tried to avoid another jet that flew too close, they lost control and crashed. Both men were killed.

Alan B. Shepard: First American to Space

When the Soviet Union launched Gagarin into space, it was a figurative 'kick in the pants' for NASA, which had not yet sent a person into orbit. Less than a month after Gagarin's daring flight, on 5 May 1961, Alan B. Shepard (an astronaut and Navy pilot) took a sub-orbital ride into space aboard a Mercury Redstone rocket, commanding the Freedom 7 Mercury capsule. His mission was not an orbital one; planners wanted to test his ability to undergo the tremendous G forces at lift-off and the rigors of atmospheric re-entry. The Freedom 7 flight therefore reached space for long enough to allow Shepard to test the attitude control systems and the retro-rockets and carry out some quick observation of Earth's surface. The flight lasted just over fifteen minutes from launch to splashdown.

Alan Shepard was a competitive man, described as incredibly mission-oriented and familiar with all the details of his spacecraft. After his first mission he was also slated to fly on a Gemini mission, but a diagnosis of Meniere's disease (an inner-ear disorder) grounded him for several years. That didn't stop him from working with NASA on other missions, however. He became Chief of the

Astronaut Office, where he worked until a surgical cure for his condition was applied.

Like most other astronauts, Alan Shepard was simply a consummate professional focused on getting the job done. For all that, he had a playful side, which the world saw at the end of his second mission, commanding Apollo 14. With his crewmate Edgar Mitchell, they collected Moon rocks at Fra Mauro base (where they landed), performed scientific experiments and drove the Lunar Roving Vehicle ('moon buggy') around. At one point Shepard, who loved to play golf, brought out a specially but covertly made 6-iron golf club and shanked two balls off in the distance and joked about hitting them 'miles and miles and miles'.

Alan Shepard retired from NASA and naval service in 1974, with the rank of rear admiral, as a hero with many decorations, including the Congressional Medal of Honor, NASA's Distinguished Service Medal and the Navy Distinguished Service Medal. He served on corporate boards and worked on a book about his accomplishments. Alan Shepard died from complications of leukaemia on 21 July 1988.

John H. Glenn, First American to Orbit Earth

The Soviet accomplishment of putting men into space placed significant pressure on the US to put their own man in orbit. The person they chose, John Glenn, was a member of the first group of astronauts selected by NASA, and a trained military pilot for the Marines. He became one of the most widely recognised public figures of the Space Age, but his fame stretched far beyond his orbital experiences. His full career included being an aviator, astronaut, businessman, senator and educator.

Glenn fought in the South Pacific in the Second World War and then in the Korean War before becoming a test pilot. Interestingly, he spent time flying with the famous aviator Charles Lindbergh during the Second World War, but never spoke much about it until later in life. Glenn was chosen in 1959 as an astronaut member of the 'Mercury Seven'. His first flight to space was aboard Friendship 7 on 20 February 1962, which orbited Earth three times before splashing down in the Atlantic Ocean. Glenn was awarded many honours in his life, including the Congressional Space Medal of Honor and the NASA Distinguished Service Medal.

Upon his retirement from NASA and the Marines, Glenn embarked on a political career. He won election to the US Senate representing the state of Ohio in 1974. A lifelong Democrat, he was nominated for vice-president three times and also ran for president in 1983. He eventually retired from politics in 1999, but not before flying aboard space shuttle *Discovery* on 29 October 1998. At the age of seventy-seven he became the oldest astronaut ever to fly in space. His work on board focused on the effects of space flight on the aged. He received the Presidential Medal of Freedom in 2012, and became an outspoken critic of the shuttle program's eventual demise. John Glenn died on 8 December 2016 after several years of declining health.

Valentina Tereshkova: First Woman to Space

On 16 June 1963, the Soviets accomplished another major goal in human space flight when they sent Valentina Tereshkova as the first woman to space. She originally worked in a textile factory and liked to skydive. Her official biographies say she made more 150 jumps and was part of a club while she was in school. She was impressed by the flights of Yuri Gagarin and Gherman Titov (the second Soviet man in space), and immediately volunteered to be a cosmonaut.

She was selected for training in 1961 and inducted as a civilian into the Air Force. Her record-breaking flight was aboard Vostok 6, on a mission that lasted three days and forty-five orbits around Earth. Like every astronaut before and after her, Tereshkova marvelled at the beauty of the planet as she passed overhead its oceans and continents.

After her record-breaking experience, Tereshkova, like Gagarin, was showered with honours, including the Order of Lenin, and Hero of the Soviet Union. She continued to work with the Soviet space programme as an engineer, but eventually left the air force at the rank of major-general to enter politics. She has served in the Russian Duma, and is admired throughout the world for her accomplishments. She remains interested in space flight, and has said she'd go to Mars, even if it was a one-way trip.

Alexei Leonov: First Spacewalk

Cosmonaut Alexei Leonov was one of the proud cadre of Soviet cosmonauts who were working to win the Space Race against the US. He flew in early Vostok missions and, had it not been for problems with the Soviet N1 rockets and eventual cancellation of the Soviet Union's crewed lunar missions, he might well have been the first man to walk on the Moon. History chose differently, and he remains to this day known for being the first man to step outside a spacecraft on orbit.

Leonov was born and raised in Siberia and always had a talent for art. He was attracted to flight from an early age, and learned enough about aviation to join a flight school directly out of high school. He became a fighter pilot before joining the cosmonaut corps in 1960. During his first mission, he became the first man to leave a spacecraft and perform what was later to become known as an 'extravehicular activity'. It was a daring and incredibly dangerous process, and he ran into a problem trying to get back into the spacecraft when his spacesuit overinflated and he couldn't get back in through the hatch. He managed to purge enough air so he could re-enter the capsule, and then had to wrestle with a hatch to seal the spacecraft.

The mission problems continued when the capsule parachuted to ground more than 161 kilometres off-course, and the crew had to wait two days for rescue. Leonov flew a second time in 1975, when he participated in the joint Apollo–Soyuz mission that lasted just over five days. After his return, he took over directorship of the Cosmonaut Training Center until his retirement in 1991. Leonov became a bank vice-president, and was awarded many honours, including Hero of the Soviet Union and the Order of Lenin.

Neil Armstrong: The Reluctant Hero Who First Set Foot on the Moon

The man who 'won' the Space Race for the US was a quiet, self-effacing person who, like many other astronauts and cosmonauts, learned to love flight as a small child. In high school he learned to fly even before he learned to drive. He studied aeronautical engineering at Purdue University and upon graduation was called up to the navy in 1949, where he began military flight training. He became a jet pilot in 1951, and served in the Korean War from 1951 to 1952. Armstrong worked as a test and research pilot, flying jet aircraft and paraglider research.

By 1960, Armstrong was a pilot consultant for a space plane called the X-20 Dyna-Soar, and selected as a pilot-engineer to test the aircraft once it was ready. Not long after that, he applied to NASA to become part of the second group of astronauts, and selected as one of the 'New Nine' in 1962 for the Gemini and Apollo programs. He also tested lunar landing vehicles, leading to a narrow escape from death when he ejected from a test vehicle just before it crashed. Fellow astronauts said he went back to work as if nothing had happened, giving him the reputation of a cool, calm commander. Neil Armstrong's most famous mission was Apollo 11, as commander of the first human mission to land on the lunar surface. He was the first person to ever set foot on another world, uttering the famous words, 'That's one small step for [a] man, one giant leap for mankind.'

It was one momentous step in a momentous life for Armstrong. After Apollo 11, he served NASA as investigator for two mishaps (Apollo 13 and the Challenger explosion). He retired from NASA in 1971 and went into private business, travelled the world and, in general, strove to live a private life away from the limelight. He was known as a hero, albeit a reluctant one. He died from complications of surgery on 25 August 2012, and was buried at sea a few weeks later.

Edwin Eugene 'Buzz' Aldrin, Jr: From the Moon to Mars

The second man to ever set foot on the Moon remains one of the most visible of the early Space Race astronauts. Edwin Eugene Aldrin Jr, nicknamed 'Buzz' from an early age, was a career military man who got his college education at West Point. He served in Korea as a jet fighter pilot, went on to teach at Nellis Air Force base, did a stint at the US Air Force Academy, and was a flight commander at Bitburg in Germany. Aldrin earned a doctorate from Massachusetts Institute of Technology in 1963, and then applied to work at NASA. His eventual selection as an astronaut led him to pilot Gemini 12, where he spent more than five hours on extravehicular activities (EVAs). Aldrin next joined the Apollo programme and was chosen for the first lunar landing, along with Neil Armstrong. He became the second person to walk on the Moon.

Aldrin left NASA in 1971 and went back to work for the Air Force at Edwards Air Force Base for a year before retiring from active duty. He has experienced significant personal problems, including bouts with depression and alcoholism, and over the years has become an advocate for fellow sufferers to get help. In 2009, he published *Magnificent Desolation* about his experiences.

Buzz Aldrin (he changed it legally to 'Buzz' in 1988) remains a tireless and flamboyant advocate for space exploration, particularly missions to Mars. In 2017, at the age of eighty-seven, he became the oldest person to fly with the Air Force Blue Angels team. Buzz has travelled the world speaking about future missions, celebrating space and science.

The Shuttle Era
John W. Young and Robert L. Crippen: The First Shuttle Pilots
Every test pilot knows a new aircraft is a collection of discoveries waiting to happen. This was particularly true of the first space shuttle, *Columbia*. It took space flight in a new direction, based on the use of wings and rockets. For its first flight on 12 April 1981, NASA selected flight-hardened veterans Captain John Young and Commander Robert Crippen to take *Columbia* on its maiden voyage.

Young's space flight career began in 1965, when he went to space aboard Gemini 3 with Gus Grissom as commander. From there, he went on to fly a second Gemini mission before joining the Apollo programme and orbiting the Moon during Apollo 10 (the final test mission before the momentous first landing in 1969). Young next commanded the Apollo 16 mission, becoming the ninth man to walk on the Moon. After his two shuttle missions, Young headed up the Astronaut Office until his retirement in 2004. Fans of the IMAX movie *Hail Columbia!* often recall his proud walk-around of the orbiter with Crippen, inspecting the machine they had just taken for 37 orbits around the planet.

Robert 'Bob' Crippen, known as 'Crip' to his teammates, was selected for the honour of *Columbia*'s first flight due to his extensive experience as a test pilot and engineer. He was originally selected for the US Department of Defense's Manned Orbiting Lab (MOL), and applied to NASA when that programme was cancelled. A former navy aviator, he served as a fighter pilot before heading back to school to study aerospace engineering. At NASA, Crippen worked as support crew for multiple Skylab missions, and the Apollo–Soyuz mission. After the first shuttle test flight, he commanded several other shuttle missions, including astronaut Sally Ride's first flight and a satellite repair mission. After the *Challenger* disaster, he worked as Deputy Director of the Kennedy Space Center. Crippen then went on to direct the space shuttle programme from NASA headquarters, before returning as director at KSC, and eventually retired from NASA in 1995.

Guion Bluford: The First African American in Space

For many years, the NASA astronaut corps was very much a white, male domain. That changed when the first women were recruited, as well as the first astronauts of colour. The first African American astronaut was Robert H. Lawrence, who was killed in a fighter jet trainer crash in 1967. Several others also served, but Guion Bluford was the first to fly in space, on 30 August 1983. Like many other astronaut candidates he came from a fighter pilot background, serving in the US Air Force (rising to become a colonel before he retired). He studied aerospace engineering and got a Ph.D. in the subject from the Air Force Institute of Technology. Bluford applied to NASA and joined the astronaut corps in 1979. He flew in four missions during his NASA career, working with payloads and doing biophysical testing on orbit. Since leaving NASA, Bluford has worked at a number of aerospace-related jobs.

John M. Grunsfeld: Hubble Hugger #1

The launch and deployment of the Hubble Space Telescope was one of the major achievements of the space shuttle era. After the discovery of spherical aberration in the telescope's mirror, NASA launched five refurbishment missions to correct the optics, replace instruments, and keep the observatory in working condition. Astronaut and astrophysicist John M. Grunsfeld was there for three of those complex missions, earning the informal designation 'Hubble Hugger #1'. Like many other astronauts of his generation, Grunsfeld grew up watching the Apollo missions and as a young child decided he wanted to be one of those people who went to space. He was highly interested in astronomy and spent time at Chicago's Adler Planetarium learning about stars and planets. It was a logical fit, since his grandfather, Ernest Grunsfeld Jr, was the chief architect for the famous facility.

John Grunsfeld is a graduate of MIT, and got his Ph.D. in physics from the University of Chicago. His main research interests have been in cosmic-ray physics (with a focus on x-ray and gamma-ray astronomy), and he has worked with NASA to develop new detectors for astronomical observations. He's also a mountaineer, and has climbed Denali, the highest peak in North America, as part of a study of body temperatures at high altitudes.

After his successful Hubble servicing missions, plus two other trips on the shuttle to do experiments on the ASTRO observatory, dock with the Mir space station and bring the Spacehab unit to space, Grunsfeld worked as NASA's chief scientist for a year. In 2010, he took on the job of deputy director of the Space Telescope Science Institute, before returning to NASA as associate administrator for the agency's Science Mission director. Grunsfeld officially retired from the agency in 2016 and is pursuing research interests in cooperation with NASA.

Sally Ride: First US Woman to Space

It took NASA two decades after Valentina Tereshkova's first flight (and the follow-up flight of Svetlana Savitskaya in 1982) to send a woman to space from the US. The astronaut chosen, Dr Sally Ride, was an intensely brilliant physicist who was equally at ease in a science lab as she was in a jet trainer and space shuttle. Her first mission was aboard *Challenger* and occurred on 18 June 1983.

Sally Ride was interested in science from childhood and had a passion for tennis, which she once said she could have pursued professionally. Ride studied physics and English, and went on to get a master's and Ph.D. in physics from Stanford University in California. She applied to the astronaut programme and was accepted in 1978. As NASA's first woman to go to space, Ride faced many questions about whether or not she could do the job of an astronaut, despite her academic qualifications and stellar performance in NASA astronaut training. Some of this resistance came from other members of the astronaut corps, as well as the press and even the technical engineering teams. Ride fielded questions about how she'd go to the bathroom in space, or tend to her monthly cycle; like other women in male-dominated professions, she dealt with these queries in a very straightforward manner and got on with the job she had set out to do.

Sally Ride served 343 hours in space aboard two missions, where she worked to deploy satellites and was the first woman to use the Canadarm robot arm (which she helped develop). In 1986, following the *Challenger* disaster, she joined the Rogers Commission investigating the cause of the accident. Her work led to the discovery of the O-ring problems that caused the explosion. In 2003, Dr Ride worked on the Columbia Accident Investigation Board, as well. She stayed on at NASA, working on strategic planning, retiring from the agency in 1987. After that Dr Ride worked at Stanford University as a policy analyst, and then went on to teach physics at the University of California in San Diego. She founded the Sally Ride Science group, which was focused on bringing science programs and publications to school classrooms, and emphasised bringing more girls into the sciences. Sally Ride died on 23 July 2013 from pancreatic cancer.

Astronauts of the ISS Era
Chris A. Hadfield: Canadian Astronaut Corps

Canada originally selected twelve astronauts to fly on space shuttles and the International Space Station. Among them was Chris Hadfield, a mission specialist on board two shuttle missions and the first Canadian to command an ISS expedition. During his last time on board the station he maintained an extensive social media presence, tweeting daily, making Facebook entries and putting up YouTube videos of his experiences. His best-known social media coup was during Expedition 35, when he released a music video of his cover of David Bowie's 'Space Oddity'. His album *Space Sessions: Music from a Tin Can* was released in October 2015, consisting of songs he performed while in orbit in the ISS.

Hadfield was born in Ontario, Canada, and grew up on a farm. In high school he learned to fly a glider in the Royal Canadian Air Cadets. He joined the Canadian Armed Forces after high school and studied mechanical engineering at Royal Military College. He served twenty-five years in his country's military. Hadfield also took flight training to become a tactical fighter pilot. His training took him to US Air Force test pilot school in California. He was eventually qualified to fly more than seventy different aircraft.

After graduating with a master's degree in aviation systems from University of Tennessee Space Institute, Hadfield applied to become one of Canada's team of astronauts, and went to work on behalf of the Canadian Space Agency at the NASA Johnson Space Center. Over the years, he worked in astronaut training, crew support and in space. He remained a civilian astronaut until his retirement in 2013. He then served as a professor in aviation sciences at the University of Waterloo and has written a book about his space experiences called *An Astronaut's Guide to Life on Earth*.

Samantha Cristoforetti: Italy's First Woman in Space

Star Trek fans were thrilled when Italian Air Force pilot and ESA astronaut Samantha Cristoforetti sent a picture of herself dressed in a Star Fleet uniform while working aboard the ISS. Her enthusiastic interest in the science fiction series endeared her to a generation of students who participated in her on-orbit programme *Mission X: Train Like an Astronaut*, at the same time that she was carrying out her 200-day mission on board the station. Like Chris Hadfield, Cristoforetti is very social media-savvy, and has frequently tweeted during her mission with her followers. While aboard the station, she used a specially made coffee machine (dubbed ISSpresso) to make the first espresso coffee in space, and tweeted a picture of herself enjoying the first sips.

Samantha Cristoforetti earned a degree in aeronautical sciences at the University of Naples. She then studied mechanical engineering at the Technische Universität Munich, Germany, where she earned a master's degree and specialised in aerospace propulsion and lightweight structures. She attended the Italian Air Force Academy from 2001 to 2005, and then was based at Sheppard Air Force Base in Texas. After completing the Euro-NATO Joint Jet Pilot Training, she became a fighter pilot and was assigned to the 132nd Squadron, 51st Bomber Wing, based in Istrana, Italy. From there she took other fighter pilot training, and gained the rank of captain in the Italian Air Force. Samantha joined ESA as an astronaut in 2009 and was selected to fly aboard the ISS in 2014, where she conducted experiments to study the effects of prolonged stays in space in low Earth orbit, with applications toward future lunar and Mars missions.

Timothy N. Peake: The First British ESA Astronaut

When Major Timothy N. 'Tim' Peake returned to Earth from his six-month tour aboard the ISS, he had travelled 3,000 times around Earth, participated in the 2016 London Marathon (while running on a treadmill in space) and broadcast New Year's greetings from space to his compatriots on the ground. Altogether, Major Peake served for six months aboard ISS. He participated in a spacewalk to repair the station's power supply, and performed a 'telepresence' experiment driving a rover across a simulated Mars terrain.

Tim Peake joined the ESA astronaut corps in May 2009, after a career as in the British Army. He served in Northern Ireland, the former Yugoslavia, Kenya and Canada, and qualified as a flight instructor in 1998. He then served as a test pilot, and after his retirement from the army had logged more than 3,000 hours of flight time on thirty different types of helicopters and fixed-wing aircraft.

Before he went to space, Peake spent time underwater as part of NEEMO – NASA Extreme Environment Mission Operations. He and his teammates lived twelve days about 60 feet underwater, developing tools and procedures for future asteroid landings. The underwater 'aquanauts' practised spacewalking techniques and tested life support systems very similar to those being used in space. Major Peake continues to work as an astronaut, and expects to return to space on another mission.

Yang Liwei: China's First Astronaut

As China ramps up its plans for extended missions in space, its astronaut crew is learning to live and work there. Its first astronaut is Yang Liwei, a highly qualified fighter pilot who was born and raised in a farming region in Liaoning province. He enlisted in the Chinese People's Liberation Army in 1983, and then began aviation training. He ultimately became a fighter pilot, clocking more than 1,300 hours of flight time. Yang was accepted as an astronaut in 1998, and took his first flight to space in 2003. He was one of a class of thirteen candidates and flew the first manned mission in Chinese history. He orbited the planet fourteen times before returning to a soft landing in central Inner Mongolia. During his flight, Yang reported hearing a 'knocking' noise in space, most likely caused by expansion or contraction of materials inside his capsule.

Liwei's flight aboard the Shenzhou 5 spacecraft was extensively publicised in his country, with the government citing it as a great step forward and a triumph for science and technology in China. He made extended tours of the country, and his name began appearing on everyday products as a tribute. He's described as being a very humble man, somewhat at odds with the adulation he has received from his country and its citizens.

Liu Yang: First Female Chinese Astronaut

As a child, Liu Yang didn't have space in her sights – at least not at first. She wanted to be a lawyer, or maybe a bus conductor. That didn't last long, and as a young adult she entered the People's Liberation Army Air Force Aviation College and eventually qualified to become a pilot. She rose to the rank of major and amassed 1,680 hours of flight experience before entering astronaut training.

Yang was selected to fly aboard Shenzhou 9, which was the first manned mission to the Tiangong 1 space station. On 16 June 2012, she became the first female Chinese astronaut to go into space. It was, coincidentally, the same date on which the first woman ever to fly to space, Valentina Tereshkova, took her first ride forty-nine years earlier. Most of Yang's tasks on her first mission were related to space medicine and effects of space flight on the human body.

Remembering the Lost

Astronauts take on many risks when they sign on to become space travellers. Anyone who flies to space has to be cognisant of the dangers and, like people whose jobs entail risk (police, firefighters, military personnel, pilots and so on), has come to terms with the chance of dying on the job. At least thirty-three astronauts and cosmonauts have died while serving their nations' space interests, nineteen of them while on missions.

The Apollo 1 Fire

The Apollo 1 fire happened on 27 January 1967. The three astronauts in the capsule were doing a full-launch rehearsal for their upcoming flight. Command Pilot Virgil I. 'Gus' Grissom, Senior Pilot Edward H. White Jr and Pilot Roger B. Chaffee were killed when the capsule caught fire. All three were experienced pilots and highly trained astronauts, and would surely have been on missions to the Moon had they survived. Their loss stunned the US space establishment and set the development of the Apollo programme back by twenty months while NASA investigated the cause of the fire.

The Soyuz 1 Accident

The Soviet Union lost at least four cosmonauts while on missions. The first accident occurred during the flight of Soyuz 1. Cosmonaut Vladimir Komarov was an accomplished pilot who flew on Voskhod 1 in 1964. Early in his career he worked on spacecraft design, particularly on the Voskhod missions. He was then selected as part of the cadre of cosmonauts for the Soyuz project, along with Yuri Gagarin and Alexei Leonov. He piloted the first test flight of the Soyuz 1 capsule on 23 April 1967. He and Gagarin (his backup pilot) knew there were safety issues with the capsule, and Komarov is said to have volunteered to fly first to protect the more famous Gagarin.

The flight had multiple problems, starting with a malfunctioning solar panel, and failure of the automatic stabilisation system. After eighteen orbits, controllers decided to bring the craft home. Upon re-entry, the parachute system malfunctioned. The spacecraft crashed to the ground, crushing Komarov inside and setting the retrorockets to fire. The loss of Komarov in the Soyuz 1 tragedy delayed the Soviet space programme for more than a year as the engineers tried to solve the problems that led to the accident.

Soyuz 11

While other astronauts and cosmonauts have died on launch or re-entry, three Soviet cosmonauts are the only ones (so far) to have died in space. They were Georgi Dobrovolksi, Viktor Patsayev and Vladislav Volkov. They launched to the Salyut 1 space station on 7 June 1971, successfully docked with it and spent three weeks working on board. On 30 June, the three men boarded the Soyuz capsule to return to Earth. Upon separating from the station, a cabin vent was somehow jostled open and it allowed the air to leak out. The cosmonauts died shortly thereafter, but their deaths were not discovered until the capsule had landed in Kazakhstan and recovery crews found their bodies inside. It was the first flight for Dobrolsky and Patsayev, and the second for flight engineer Volkov. All three were posthumously awarded Hero of the Soviet Union medals.

This disaster uncovered severe problems with the capsule's design. The failing valve was located between the descent module of the spacecraft and the orbital module. When the two sections separated, explosive bolts misfired and the valve opened at an altitude of 168 kilometres. The men weren't wearing space suits, which might have saved their lives. The incident led to a revamp of the capsule to carry only two cosmonauts, leaving room for each to wear a space suit for launch and re-entry. Later versions of the Soyuz capsule were redesigned to allow for three passengers wearing lightweight space suits.

The *Challenger* Catastrophe

One of the most devastating space tragedies of the Space Age occurred on 28 January 1986 in full view of the world via television and satellite coverage. The *Challenger* space shuttle set off carrying seven astronauts, offering an amazing cross-section of Americans. Christa McAuliffe was an experienced teacher from Concord, New Hampshire. Ron McNair was the second African American to fly in space, and had a Ph.D. in physics, working with lasers. He was also a saxophonist, and had planned to play an original musical composition by Jean-Michel Jarre while on orbit. The piece, called *Rendez-vous* (and nicknamed 'Ron's Piece') would have been part of a live concert feed between the shuttle and the ground. Judith Resnik was the second US female astronaut to fly and had an extensive background as an electrical engineer. Commander Scobee came from a solid military background and served in Vietnam as a fighter pilot. He piloted *Challenger* once before his final mission. Michael Smith, Scobee's second-in-command, had a similar military background, and had extensive experience with more than two dozen types of aircraft. Ellison Onizuka, who was on his second shuttle flight, was the first Asian American to fly for NASA. He was originally from Hawai'i and served in the US Air Force before joining NASA. *Challenger* was his second mission. Crewmate Gregory Jarvis also served in the US Air Force and worked in private aviation before joining NASA to become an astronaut.

They were to deploy a TDRS-B tracking and data relay satellite, plus the Spartan-Halley space probe, being sent to study Comet Halley in ultraviolet light.

The shuttle was destroyed in an explosion seventy-three seconds after lift-off. The cause of the accident was traced to gases escaping from the right solid rocket booster, which damaged the booster's attachments and led to the failure of the external fuel tank. The orbiter, which was not damaged by the tank failures, was blown apart by aerodynamic forces, and pieces of it (including the intact crew compartment) fell into the Atlantic Ocean off the coast of Florida. The accident prompted a two-and-a-half-year hiatus in shuttle flights while the Rogers Commission investigated all the factors leading to the accident.

The Loss of *Columbia*

NASA's third major space catastrophe took the lives of seven astronauts when their orbiter broke up on re-entry on 1 February 2003. Commander Husband was a colonel in the US Air Force and studied mechanical engineering in college. He was an accomplished pilot and worked on advanced research projects and flight safety before his assignment to flight duties. Husband's second-in-command, William 'Willie' McCool, came up through the US Navy, and studied computer science and aeronautical engineering. He became a naval aviator, flying more than two dozen types of aircraft with over 2,800 hours of flight time. He joined NASA in 1996, and the 2003 *Columbia* mission was his first flight. Physicist and US Air Force officer Michael Anderson was in charge of science experiments on the ill-fated mission. He had degrees in physics and astronomy, and joined NASA in 1994, where he flew aboard *Endeavour* in 1998. Crewmate Colonel Ilan Ramon was the first Israeli astronaut to fly, coming to the agency from a career as a fighter pilot in his country's air force. Mission specialist Kalpana Chawla had one previous mission on *Columbia* to her credit. She joined NASA in 1995, the first Indian American to fly in space. Mission specialist David Brown was a navy pilot and flight surgeon. He was an accomplished gymnast and almost became a circus performer upon leaving college. Brown joined NASA in 1996 and became a mission specialist. Laurel Clark, who was also a physician and navy captain, served as a flight surgeon and undersea doctor during her navy career. She too joined NASA in 1996, and *Columbia* was her first space flight.

On the day of the disaster, *Columbia* was its way home after a successful two-and-a-half-week mission. During the 16 January 2003 launch, observers noticed a piece of debris strike the orbiter's left wing. Concerns were raised about whether or not it had damaged the wing, and eventually it was decided the evidence for damage was not strong and the shuttle was given the 'ok' to return home. During descent, the wing-edge temperatures rose due to the heat of atmospheric friction, and this led to a cascade of failures in the hydraulic and control systems. The orbiter broke up high in the atmosphere, and pieces of debris flying from what was *Columbia* could be seen from the ground. This whole sequence of events occurred as the shuttle was coming in for its approach and landing at the Kennedy Space Center. NASA suspended flight operations while the investigation teams gathered debris and worked out the timeline of the orbiter's destruction.

There have been many posthumous awards and memorials to the astronauts and cosmonauts who have died on their missions. Research buildings, science centres, monuments and highways bear their names. Asteroids, as well as surface features on several solar system objects, bear the names of the lost, including the *Columbia* Hills on Mars, craters on the Moon, and a mountain range on Pluto.

Astronauts of the Future

The term 'astronaut' means 'space sailor', which is a fitting description of the men and women who live and work in space. Even though the first missions they've flown have been limited to Earth and the Moon, in the future astronauts will sail to other worlds. There are those among us today who will be taking the first steps on Mars. Others will take pioneers to the Moon, and – in the far distant future – will make the leap to the worlds of the outer solar system. No matter where they live, or which nations they call home here on Earth, astronauts and cosmonauts belong in space, taking the rest of us to the new frontier.

4

Global Steps to Space: A World of Space Agencies and Institutes

Our two greatest problems are gravity and paperwork. We can lick gravity, but sometimes the paperwork is overwhelming.

Wernher von Braun

Eventually private enterprise will be able to send people into orbit, but I suspect initially it's going to have to be with NASA's help.

Sally Ride

I feel very strongly that SpaceX would not have been able to get started, nor would have made the progress that we have, without the help of NASA.

Elon Musk

We live in a global Space Age. Hundreds of thousands of people around the world work for agencies and research institutes directly involved with space exploration. In the early decades of space exploration, individual agencies such as NASA and the design bureaus of the Soviet Union worked independently of each other. As other countries got interested in space they formed their own institutions for research and development, and today many countries are involved in individual and cooperative space efforts.

As with the United States and Soviet Union in the decade after the Second World War, many governments in the world had a deep interest in going to space. This was at a time when some countries were developing their own rocket programs. Today, more than seventy governmental space agencies exist, including the cooperative European Space Agency, which unites twenty-two member states plus advisory states in a common effort. Thirteen agencies around the world have the ability to launch rockets and payloads to space. As of mid-2017 only Russia and China have human launch capability, with NASA relying on Roscomos to send astronauts to the ISS.

In addition to space access and scientific research, countries have always been interested in controlling their satellite communications. This has led to tremendous growth in the global communications and reconnaissance satellite industries. In this chapter, we'll look at the main agency players on the space stage.

NACA and NASA

Before the Space Age ramped up, space technology and aeronautics in the US were under the aegis of the military as well as the National Advisory Committee for Aeronautics (NACA). It was formed in 1915 as an independent agency to coordinate aeronautical research in the US. The first chairman was Brigadier General George Scriven, Chief of the Army Signal Corps. He had a committee of twelve advisors taken from government, industry and military sectors that met a few times each year to discuss new efforts in aeronautics, but as flight technology grew NACA's role expanded. In 1920 it founded the Langley Aeronautical Laboratory in Virginia, which heralded an expansion in the fledgling agency's mission and personnel. The main mission was to conduct flight tests in specially designed wind tunnels built by the agency. The advances from these tests contributed to more streamlined aircraft.

The Second World War forced an accelerated pace of aircraft development, and NACA rose to the challenge, working with the military and various companies (McDonnell, Douglas, Hughes, Lockheed and others) on advanced wing and propeller designs. This required more advanced testing facilities. The Ames Aeronautical Laboratory was built in 1940, followed by the Aircraft Engine Research Laboratory in 1941.

At the end of the war, NACA pivoted its attention toward designs for more advanced supersonic aircraft. New labs in California sprang up to accommodate flight testing, and NACA began collaborating with the private company Bell Aircraft, which manufactured fighter planes and helicopters. The Bell X-1, an experimental rocket plane, was eventually used to break the sound barrier for the first time. This was the flight made famous by pilot Captain Charles 'Chuck' Yeager in 1947. The design of the X-1 radically changed the look and feel of high-speed flight.

NACA entered the 1950s with a focus on rocket technologies that would primarily deliver warheads to enemy targets. Other agencies, such as the Army Ballistic Missile Agency and the Navy Research Labs, were moving ahead with rocket developments and top-secret plans for scientific and reconnaissance satellites, respectively. The Cold War was in full swing by the mid-1950s, and the committee continued to build testing sites for rocket-propelled planes and missile warheads, such as a test range at Wallops Island in Virginia.

Around the same time, NACA was tasked with looking into the technology and techniques that would eventually be used to launch payloads to space. This included sending humans and bringing them back to the ground. The trip would require a heavily shielded spacecraft that could survive the rigors of launch and a return through the atmosphere – no easy task. It required the development of shields, hardened electronics, specialised cockpit instruments, and other technology. In addition, there had to be some way to communicate with the spacecraft, and so NACA was charged with research and development into that area, too.

The launch of Sputnik 1 pushed NACA squarely into a hotly contested political realm. Suddenly, the race to space was on. The fact that the Soviets had leapt to space first gave the public the impression that the US was somehow 'behind'. That was one reason why US President Eisenhower suggested the creation of a new agency focused on aerospace activities, scientific research, and technical advancement in space. The decision was made to make the US space programme a civilian one. On 1 October 1958 NASA was chartered, absorbing NACA and its test facilities and offices. The new agency was aimed at attaining and maintaining US primacy in space. It would take control of satellite and unmanned missions, and continue NACA's emphasis on research into aeronautics designs.

NASA also acquired test units and science labs from the army – including the facilities and personnel at the Jet Propulsion

Laboratory in Pasadena, California and the Redstone Arsenal in Huntsville, Alabama, where Wernher von Braun and his team were developing the successors to the V-2 rockets they'd built in Germany. It was a monumental undertaking to unite all these groups under the NASA umbrella to focus on going to the Moon. After the lunar missions, the agency began designing and deploying space stations and the shuttles, plus a myriad of planetary probes. Ultimately, it shepherded missions to all the known planets, asteroids and comets.

NASA's Structure

NASA is a federal agency, part of the US executive branch whose leadership is appointed by and answers to the President. It is headquartered in Washington, D.C., and there are ten NASA field centres, each with its own areas of specialisation. There are also several research groups: Goddard Institute for Space Studies, the Independent Verification and Validation Facility, the Michoud Assembly Facility, the NASA Safety Center, NASA Engineering and Safety Center, the NASA Shared Service Center and the Wallops Flight Facility. In addition, NASA works with researchers at major universities around the world, funding graduate and undergraduate programs and study teams. The agency has over 17,000 employees.

Ames Research Center

The Ames Research Center at Moffett Field, California, was originally part of NACA and was taken into NASA when the parent agency formed. It's named after Joseph S. Ames, who was a physicist and member of NACA. It performed wind-tunnel tests as part of NACA's original mission. Today, research at Ames includes robotic lunar exploration, supercomputing, thermal protective materials, astrobiology (the study of the origin, evolution, distribution and future of life in the universe) and airborne astronomy. The Stratospheric Observatory for Infrared Astronomy (SOFIA) is administered at Ames and is a joint project with the Universities Space Research Association and the German SOFIA Institute at the University of Stuttgart. SOFIA is a modified Boeing 747SP aircraft has a 2.5-metre telescope mounted so it can observe the sky as the plane flies high in the atmosphere. Ames has the world's largest wind-tunnel test facility (big enough to test a 737 airliner), and maintains the national Full-Scale Aerodynamics Complex on its grounds.

The centre also hosts mission operations for the Kepler space telescope, the Lunar Crater Observing and Sensing (LCROSS) mission, and other projects.

Armstrong Flight Research Center

Edwards Air Force Base in California hosts the Armstrong Flight Research Center (formerly called the Dryden Flight Research Center), now named in honour of astronaut Neil Armstrong. It has a long history of flight test operations, including the Lunar Landing Research Vehicle used to train Apollo astronauts. This installation operates aerodynamics and aeronautics test beds using some of the most advanced aircraft known, including the F-15B jet aircraft and the Gulfstream III. The Stratospheric Observatory for Infrared Astronomy (SOFIA) is based at Armstrong.

Glenn Research Center

The Glenn Research Center (named in honour of astronaut John H. Glenn) is one of NASA's oldest centers. It is part of Lewis Field near Cleveland, Ohio, and was first created by NACA in 1942 as a laboratory for aircraft engine research and development. In recent times its attention has shifted to exploring highly energy- and fuel-efficient engine designs, as well as noise-reduction technology. Today it conducts large-scale engine tests, spacecraft propulsion studies, and zero-gravity research. In addition to its technical mission, the Glenn center provides internships for students and educators.

Goddard Space Flight Center

The Goddard Space Flight Center in Greenbelt, Maryland was established by NASA on 1 May 1959 as the agency's first space flight centre. It was named in honour of rocket pioneer Robert H. Goddard. It employs a mix of NASA technical and scientific staff as well as contractors from private industry. Goddard is a laboratory where manned and unmanned spacecraft (such as the James Webb Space Telescope) are developed, built and tested. The centre also houses control facilities for many missions, including the Earth Observing System, the Hubble Space Telescope and the

Solar Dynamics Observatory satellite, as well as Geostationary Operational Environmental Satellite (GOES) system, and others.

Jet Propulsion Laboratory

The Jet Propulsion Laboratory is located in Pasadena, California, nestled into a set of hills that once echoed with the sound of rocket motors. Today, it's a sprawling campus of offices, test chambers, and spacecraft design and manufacturing facilities.

JPL's beginnings in what was once a dusty arroyo date back to the 1930s, with three CalTech graduate students and their colleagues. Frank Malina, Qian Xuesen, Weld Arnold, Apollo Smith, Jack Parsons and Edward S. Forman – overseen by Theodore von Kármán – managed to get some army money to test rockets and eventually jet engines. Malina, Parsons, Forman and von Kármán went on to create the Aerojet Corporation, to design and create jet-assisted take-off motors. The Jet Propulsion Laboratory was formally established in 1943 on the testing grounds where the men did their first work.

The lab was transferred to NASA in 1958 and established as a planetary spacecraft centre. The lunar missions Ranger and Surveyor were built and operated by JPL scientists and technical staff. Over the years, other missions to the Moon and planets have been developed and run by JPL. Its 'mission control' areas are world-famous from TV broadcasts for the Voyager, Mars Pathfinder, Curiosity and other missions. Now managed by CalTech for NASA, JPL has an extensive educational outreach programme, particularly to museums, and provides internships and training opportunities for students.

Johnson Space Center

The Johnson Space Center (named for former US president Lyndon B. Johnson), in Houston, Texas, is 'mission control' for NASA's crewed missions. Built just after NASA was formed, it houses the Astronaut Office and training centres, the Christopher C. Kraft Jr Mission Control Center, the neutral buoyancy tanks, astronaut training classrooms and labs. When the Apollo missions brought back lunar samples, the Lunar Sample Laboratory and containment facility was created to house them. Medical researchers at Johnson study the effects of extended space flight on the human body.

Kennedy Space Center

The Kennedy Space Center (named for former President John F. Kennedy) is the most visited of NASA's facilities. It is located on Cape Canaveral in Florida (east of Orlando), where most launches have taken place since 1968. It is adjacent to the Cape Canaveral Air Force Station, where mostly military launches occur. Both facilities are surrounded by large wildlife sanctuaries.

KSC is home to Launch Complex 39, responsible for all launches of the Apollo and shuttle programs, plus the Skylab and many of the space station modules. The largest building in the world, the Vehicle Assembly Building (VAB), is also located at KSC, along with the launch control centre, and news media facilities.

A few kilometres outside the KSC gates, the Kennedy Space Center Visitor Complex offers exhibits related to space exploration, plus gift shops, simulators, a rocket 'garden' and a memorial to those lost on space-related duty. The main attractions on display at the complex are the space shuttle *Atlantis* and the last Saturn V rocket.

Langley Research Center

In the first decades of the twentieth century tensions with Germany led the US to establish aeronautics research facilities to perform advanced studies on aircraft and aeronautics. The first of these was the Langley Memorial Aeronautical Laboratory, named for Samuel Pierpont Langley, an astronomer and aviation pioneer. The centre's main research areas were airframe and engine design, and it once housed the world's largest wind tunnel. In the 1940s, the centre began rocket research and expanded out to Wallops Island. Langley was incorporated into NASA just after the agency formed, and continues a tradition of aeronautical research. The centre was a site where astronauts trained for the Gemini rendezvous missions, and a test facility for the Apollo missions.

Today, Langley works on many diverse projects: drone management, crash safety and biofuels for jet engines, as well as earth-sensing satellites, robotic technologies, deep-space missions and the Journey to Mars Initiative.

Marshall Space Flight Center

When German rocket engineers came to the US after the Second World War, they eventually found a research home in Huntsville, Alabama. At the time it was the home of the army's Redstone Arsenal, a rocket proving ground, and eventually was the home of the Army Ballistic Missile Agency. It was a hive of missile development and testing. On 21 October 1959, the Army's missile research programme, particularly for space uses, was transferred to the newly formed NASA, and the George C. Marshall Space Flight Center was opened at Redstone Arsenal. Marshall was a former army general and author of the Marshall Plan that transferred aid to western Europe after the war.

The Marshall Space Flight Center played a major role in the Apollo, Skylab and Spacelab missions, and in constructing the Destiny module on the ISS (built by contract with Boeing). MSFC was responsible for research and development of the space shuttle's propulsion systems, working with such contractors as Aerojet Rocketdyne, Morton Thiokol and others. The space centre takes the lead role in all of NASA's launch systems, including the Space Launch System. Its space research programs are involved with zero-G experiments, as well as deep-space astronomy missions such as the Hubble Space Telescope and the Fermi Gamma Ray Space Telescope.

Although not directly connected to the Marshall Space Flight Center, the nearby US Space & Rocket Center serves as its visitor centre. It plays home to US Space Camp and contains an extensive collection of rockets and space exploration artefacts.

Stennis Space Center

NASA maintains a state-of-the-art rocket testing facility called the Stennis Space Center, located in Mississippi. It hosts national and international agencies and companies (such as Aerojet) that make use of the test beds. Stennis is near the Michoud rocket assembly factory in New Orleans, and rockets are routinely 'barged' over for testing. Stennis was built in 1961 and used first for testing the engines used in the Apollo missions. Space shuttle engines underwent testing and flight certification there. Today the centre maintains some of the most extensive rocket test stands in the world through its Engineering and Test Directorate. The official visitor centre for Stennis is located at INFINITY, a nearby science museum.

As mentioned throughout this book, NASA is a technology driver in the US, and has been involved in education and public outreach at all levels. Its charter mandates the agency to share its findings with the taxpaying public, which it does through museums, schools, webpages and various activities.

The Rise of the Soviet and Russian Space Efforts

The Soviet interest in space travel began long before the launch of Sputnik 1. That event was really the equivalent of opening the door to a busy lab in operation for many years. In chapter 1 we looked at the rise of rocket technology, which was driven by such geniuses as Robert H. Goddard in the US and Konstantin Tsiolkovsky, Hermann Oberth and Sergei Korolev in the Soviet Union in the early part of the century. The capture of German rocket technology after the Second World War put their development for both military and exploratory purposes on the fast track. In the Soviet Union, space exploration was not relegated to one agency as it eventually was in the US.

The Soviet space programme was divided into military and scientific efforts, using a standardised process for developing missiles, rockets, spacecraft and probes. Beginning in the 1930s, multiple design bureaus – such as Korolev's OKB-1 group (OKB stands for 'Osoboe Konstruktorskoye Buro', or Special Experimental Design Bureau) – and institutes were involved in Soviet space efforts. There were multiple such OKB institutions, many working on aircraft. The space-related bureaus were responsible for systems or sub-systems of rockets and spacecraft. Korolev's bureau was tasked with the development of the Sputnik and Vostok flights, which were major initiatives beginning in the late 1950s and into the 1960s.

The competing design bureaus worked on both civilian and military aspects of the Soviet space programme for many years, with the push to the Moon as a major driver. They have also designed and built communications satellites. Throughout the 1960s and 1970s, OKB-1 continued to develop the needed hardware for the lunar missions. (Today, OKB-1 is known as OAO S. P. Korolev Rocket and Space Corporation Energia, or Energia for short.)

Once it was clear that the Americans would win the Moon race, the Soviets turned their attention to space stations – the Salyut and Mir programs being most familiar – and planetary probes. Soviet spacecraft either passed by or landed on the Moon, Venus and Mars. The Soviet space programme flew the first men and women in space, sent the first balloons to Venus, had two probes fly by Halley's Comet in 1986, built the first space station with permanent crews and set cosmonaut records for sojourns in space.

When the Soviet Union dissolved in late 1991, the newly formed Russian Federation consolidated its space exploration activities to form the Russian Space Agency. That led to some substantial jostling for power positions among the bureaus, among them Energia, which controlled the Mir space station. The corporation's Soyuz spacecraft are (as of 2017) the only way astronauts can get to and from the ISS.

Under the Russian Federation the space agency's budget suffered for a time, but eventually the country signed on as a partner to the ISS project. In 2013 the space industry in Russia went through major reforms, and the Russian Space Agency was dissolved. Responsibility for space missions was placed under the aegis of Roscosmos State Corporation in 2015. It merged with the United Rocket and Space Corporation, a Russian government entity. This new state corporation is now in charge of all international efforts in space involving Russian hardware and crews.

Space Centres

Roscosmos operates a space centre at Baikonur, a satellite and ICBM launch facility at Plesetsk, and also uses the French Korou Launchpad in Guiana.

The main Russian space launch centre is Baikonur Cosmodrome, a large spaceport in Kazakhstan with nine launch sites, testing areas for ICBMs, assembly labs, centres for launch vehicle processing and preparation, office complexes, two airfields and associated infrastructure for communications, electricity and transportation. It began in the Soviet era as Scientific Test Range #5, and is the site of many of the country's launches. It is currently where all astronauts going to the ISS prepare for their trips. During the short-lived Buran programme, Baikonur was its home base.

Closer to Moscow, Roscosmos is developing the Vostochny civilian cosmodrome. Its first launch took place on 28 April 2016. but construction difficulties have delayed its full use indefinitely.

European Space Interests
France's National Centre for Space Studies
The European ascent to space began with a number of countries creating their own space exploration and science groups in the aftermath of the Second World War. The French formed the Laboratoire de Recherches Balistiques et Aérodynamiques in 1946 to build their own rockets using knowledge of the German V-2s. In 1958, President Charles DeGaulle ordered the creation of committees responsible for developing French space exploration expertise. That led to the Centre d'Etudes Spatiales (CNES), and the country began developing rockets for launching satellites. France built launch pads in Kourou (Guiana), and participated in the creation of the European Space Agency.

German Aerospace Centre
The Deutsches Zentrum für Luft- und Raumfahrt e.V. (DLR) is Germany's unified national centre for research into aerospace, energy, transportation and related sciences. It has only existed since 1997, but comprises agencies and research institutes that date back to the early twentieth century. The current agency's mission is focused on Earth-based exploration, environmental protection, and basic and applied research. The agency is based in Cologne, with institutes across Germany. It coordinates activities for the Columbus laboratory on the ISS, and has been heavily involved in the Mars Express, Rosetta and Dawn planetary and comet missions. DLR operates an extensive set of research aircraft, and is also involved in emissions research, and solar power tower development. DLR is a partner in the European Space Agency.

Italian Space Agency
The Agenzia Spaziale Italiana was created in 1988 as part of Italy's Ministry of Education, Universities and Research. Like most other space agencies, it had research predecessors in aerospace, including the Italian Space Commission and the Centro Ricerche Aerospaziali. These groups cooperated in the 1960s to create and launch Italy's first satellite, the San Marco 1 (at NASA Wallops). Italy went on to launch several of its own satellites and was a key player in the European Launcher Development Organisation and the European Space Research Organisation (which later became the European

Space Agency). Italy's best-known mission was the BeppoSAX x-ray satellite, but its team has also been involved in missions with NASA and the ESA, including Cassini-Huygens, INTEGRAL, Mars Express, Dawn, and Juno. ASI is part of the Ariane 5 launch consortium, and participates in Earth observation studies, sends astronauts to the ISS and helped build several of the station's modules.

Spain's Instituto Nacional de Técnica Aeroespacial

The Instituto Nacional de Técnica Aeroespacial has been around in one form or another since 1942 and encompasses both civilian and military research. It is part of a larger effort that includes naval, defence and energy technologies. It has been responsible for developing and launching low-cost reconnaissance and communications satellites.

The agency is best known for supplying the Madrid Deep Space Communication Complex and the El Arenosillo launch site.

UK Space Agency

While Britain and British companies have long been involved in aerospace research, astronomy and Earth sciences, the United Kingdom Space Agency is a relative latecomer, formed in 2010. The agency was created to improve the country's efforts in space exploration and to make strategic decisions for UK civil space activities. As part of the Department for Business, Energy and Industrial Strategy, the agency continues to pursue research opportunities. One of its most high-profile missions was 'Principia', the six-month sojourn of astronaut Timothy N. 'Tim' Peake aboard the ISS. The UK Space Agency is part of ESA.

ESA: A Unified European Space Effort

Although individual European nations continue to pursue their own interests in space exploration, many have elected to band together to form the European Space Agency. In 1962, two agencies were charged with developing the means to go to space and were not controlled by the US or the Soviet Union – the European Launch Development Organisation (ELDO) and the European Space Research Organisation (ESRO).

ELDO was a research and technical group for designing and building launch vehicles for European use. Scientists and technical staff from seven countries (Belgium, Britain, France, Germany, Italy, the Netherlands and Australia, which supplied a launch complex in Woomera). ELDO worked on the Europa launcher, which suffered launch failures and was eventually cancelled. Rocket design then became the province of the French Ariane project in 1973, now Arianespace. The group has designed and built five rocket models in the Ariane series. ESRO was a mainly scientific enterprise founded in 1964, focused on space science and telecommunications. ELDO merged with ESRO to form the European Space Agency in 1975.

ESA has launched missions to other planets, satellites for communications and surveillance, and participated in large multi-national projects such as the ISS. ESA's common aim is to be Europe's gateway to space. Its twenty-two member states are: Austria, Belgium, Czech Republic, Denmark, Estonia, Finland, France, Germany, Greece, Hungary, Ireland, Italy, Luxembourg, the Netherlands, Norway, Poland, Portugal, Romania, Spain, Sweden, Switzerland and the United Kingdom. Other countries have signed cooperative agreements with ESA, including Bulgaria, Cyprus, Malta, Latvia and Slovakia; Slovenia is an associate member and Canada has a special relationship with the agency as well.

ESA is headquartered in Paris, but like NASA, with its many space centres, has facilities around the continent. For example, the agency's astronauts and ground personnel train at the European Astronauts Centre in Cologne, Germany. It's also the home of the Space Medicine Office. The European Space Astronomy Centre (ESAC), near Madrid, Spain, houses science operations centres for ESA's planetary and astronomy spacecraft, and is host to the vast data archives they produce. The European Space Operations Centre (ESOC) in Darmstadt, Germany, handles operations for the engineering teams that control satellites and coordinate with global tracking systems of near-Earth asteroids. ESOC flight operations teams plot orbital trajectories and maintain contact with missions The European Space Research Institute (ESRIN) is the main centre for Earth observations and climate studies. It's located in Frascati, near Rome, Italy. The European Space Research and Technology Centre (ESTEC) in Noordwijk, the Netherlands, is an 'incubator' for developing space projects and technologies. It also supports a test centre for spacecraft. ECSAT, the European Centre for Space Applications and Telecommunications, Harwell, United Kingdom, functions as the business centre for telecommunications, technology applications, and aspects of climate change. It is known as UK's Space Gateway. ESA's Redu Centre in Belgium is mainly responsible for controlling and testing satellites through the ESA ground station network. Personnel also work on the Space Situational Awareness programme, which tracks man-made and natural debris that could come into contact with Earth or space-based habitats.

ESA is an important part of the political and economic engines driving European aerospace. Its annual budget is around €5 billion, which amounts to a few euros per taxpayer, and it employs just over 2,000 people across its installations.

China National Space Administration

China has long been interested in space exploration, partly in response to the American and Soviet expansion of both scientific instruments and weapons into low Earth orbit. The country pursued access to space as a way to deploy its own strategic nuclear and conventional weapons on the high frontier. That started in the late 1950s with the development of its ballistic missile programme under the auspices of a Sino-Soviet accord. Under a cooperative technology programme, the country got access to Soviet R-2 rocket technology. That lasted until 1960, when the two countries split. The Chinese began launching their own rockets in September 1960, and continued testing both military and scientific payloads.

The late 1960s and early 1970s saw China begin the development of its own human space flight programme. This proceeded in fits and starts as the country experienced political upheavals. After the death of Mao Zedong, the country continued testing ballistic missiles for both conventional and nuclear warheads. Eventually, in 1988, the Ministry of Aerospace Industry was created, and was later split into two sections: the China National Space Administration (CNAS) and the China Aerospace Science and Technology Corporation.

The CNAS launched its first astronaut to space, the aforementioned Yang Liwei, aboard the Shenzhou 5 capsule atop a Changzheng 2F rocket (part of the Long March 2 family). The flight lasted twenty-one hours and made China the third country to achieve human space flight.

In addition to the human launches, China has launched two space stations, Tiangong 1 and Tiangong 2, using the Tianzhou-1 cargo spacecraft. The first station has been de-orbited, but the second station remains operational and houses a variety of science experiments and astronomy-related observations. A third station, called Tiangong 3, is in the works and should be launched and operating in the early 2020s.

China is currently focused on lunar missions, and has sent both orbital and lander missions to the Moon's surface to scout out the terrain. Sample return missions and a possible crewed visit appear to be the next step for the Chinese space programme. As with other countries, China is also eyeing missions to Mars in the distant future. One of its most ambitious missions is Insight, the Hard X-ray Modulation Telescope, its first astronomy satellite, used to observe black holes and neutron stars.

As with other spacefaring countries, China is entering into international collaboration agreements for future exploration. The country's interest in lunar missions has led it to partner with the ESA to build a human outpost on the Moon. The Moon Village would be a test bed for exploration technologies, space tourism and lunar mining. The countries also foresee it as a development base for eventual missions to Mars and beyond. Such collaborations are currently off-limits to NASA workers due to US government prohibitions, although China has said it is open to future cooperation with the Americans. That freeze may be thawing soon, with the deployment of a joint commercial project between the Beijing Institute of Technology's School of Life Sciences and Texas-based NanoRacks, which puts a DNA science experiment aboard the ISS. Another possible outcome of a lunar village would be the construction of space-based solar power satellites used to beam energy back to Earth for China's consumption.

CSNA maintains several satellite launch centres. The country's first spaceport is at Jiuquan in the Gobi Desert. It is part of a larger 'space city', with a museum and other facilities, and is used to launch satellites and other vehicles into low and medium orbits. The first Chinese astronauts were launched from there in 2003, and another mission (Shenzhou 9) lifted off in 2012.

The Xichang Satellite Launch Center is located in Sichuan Province and is the site of most heavy-lift launches for communications and weather satellites. Many of its functions are being transferred to the

Wenchang Center, which is located in Hainan, China. Wenchang is specially situated at low latitude and is mainly used for sending the newer classes of Long March boosters to space. It is used for space station and crew launches, and the country's deep-space and planetary missions.

The Taiyuan Satellite Launch Centre has been used mainly for weather satellites and earth-science satellites, as well as ICBMs and other defensive concerns.

Control centres for Chinese missions exist in Beijing and in Xi'an. In addition, the CNES maintains a fleet of tracking ships, an extensive deep-space tracking network utilising antennas in Beijing, Shanghai, Kunming and at other sites around the world.

Japan to Space

After the chaos of the Second World War, Japan was banned from pursuing any kind of aviation or rocket development for several years. The restrictions placed on the country hindered any advances in those areas. Once the final peace treaty was signed in 1951, Japanese researchers could pursue aviation technology, and in particular the chance to build rockets. Like China, Japan watched the early years of the Space Race even as its scientists were experimenting with rocket design and launches. The early 1950s saw researchers begin construction of Japanese rockets at the University of Tokyo's Industrial Science Institute. The effort was led by Hideo Itokawa, who developed and launched the Pencil Rocket. Later he and his team built and tested the Baby and Kappa rockets.

Itokawa's work dovetailed with Japan's interest in participating in the International Geophysical Year. After a decade of research and testing, the first Japanese satellite, called Ōsumi, was sent to space on 11 February 1970, bringing Japan into the Space Age.

In the decades since then, the National Aerospace Laboratory of Japan (NAL), the National Space Development Japan (NASDA) and the Institute of Space and Astronautical Science (NSAS) all worked on various aspects of the country's space programme before merging in 2003 to form the Japan Aerospace Exploration Agency (JAXA). This organisation guides technology development, funds and directs scientific research, and in recent years has funded advanced missions to asteroids. JAXA has outlined plans for possible missions to the Moon and Mars, but those remain on the drawing board due to less than favourable budgeting priorities.

Through the years, JAXA has entered into cooperative missions with NASA and Roscosmos, and supplied the Kibo module to the ISS. Its planetary missions have included flybys of Comet P/1 Halley in 1986 by the Suisei and Sakagaki probes, the Hayabusa missions to the near-Earth asteroids 25143 Itokawa and 162173 Ryugu (sample-return projects), Nozomi to Mars, Hiten and Kaguya doing lunar exploration, and a solar sail project called Akari. Future missions include participation with ESA on the Jupiter Icy Moon Explorer (JUICE) probe. Japan is also heavily involved in space astronomy, with infrared telescopes, x-ray and radio astronomy missions, and solar observations. It has launched navigation satellites, Michibiki-1 and Michibiki-2, to work with the existing GPS system to improve the accuracy of navigational signals.

Like other space agencies, JAXA operates other institutes, such as the Earth Operations Centre in Hatoyama. The Institute of Space and Aeronautical Science (ISAS) operates all the agency's planetary missions. As a result of the Japanese government's comprehensive strategy on science, technology, and innovation, JAXA established a space exploration innovation hub. It is charged with developing research projects. To facilitate the use of spinoffs, JAXA created a New Enterprise Promotions department to support research and development on ground-based technologies, and to look for new opportunities in space (specifically aboard the ISS).

India Races to Space

India is a relative newcomer to the Space Age. Its main research agency – the Indian Space Research Organisation – was formed in 1969. Space research has been a goal of the government since the country became interested in using its own satellites for communications. Since then, the country has fielded a healthy share of well-trained scientists in astronomy, rocket and space sciences. Its vision is to do what it takes to harness space technology for national development and economic prosperity, and to use space science, research and planetary exploration as a way to do it.

ISRO built its first satellite, Aryabhata, and sent it to orbit aboard a Soviet Kosmos 3M rocket in 1975. That was followed in

1980 by the Rohini mission, the first launch aboard an Indian-built rocket. Since that time, ISRO has built more powerful rockets to launch the country's burgeoning fleet of communications satellites, earth-pointing probes and planetary missions.

ISRO undertook the deployment of the Satellite Instructional Television Experiment (SITE) in 1975, jointly with NASA. It used NASA's ATS-6 satellite to beam educational television shows to people in rural India. These programs reached 200,000 people, including 50,000 science teachers. Another project called INSAT used geosynchronous satellites for domestic communications. With these and other communications projects, India ramped up its distance learning capabilities.

ISRO maintains a deep interest in solar system science, and in 2008 sent its first lunar orbiter, called Chandrayaan. The follow-up mission, called Chandrayaan 2 includes a lunar orbiter, lander and rover. The agency's Mars Orbiter Mission (MOM), built as a low-cost technology demonstrator, made it to Mars. In early 2017, ISRO demonstrated its launch capabilities by delivering twenty satellites in a single payload.

The ISRO has its sights set on even bigger projects: human missions to space using the GSLV Mrk III heavy-lift vehicle; a reusable launch vehicle called Avatar, part of the country's Defence Research and Development Organisation; more interplanetary probes; and Aditya-L1, a mission to study the Sun from an L1 orbit. Through the Antrix Corporation, its commercial wing, ISRO hopes to garner business from other countries interested in affordable launch avenues for satellites.

Up-and-Coming Agencies and Institutes

Space is not just limited to the wealthier and larger countries and their partners. Many other countries are looking to space exploration as way to stimulate education and commerce in the science sector. It's worth taking a look at a select sampling of these smaller agencies and their accomplishments.

Israel

Currently, Israel's launch capability is used for sending satellites to orbit. The Israeli Space Agency was formed in the 1960s, originally as a research project at Tel-Aviv University. Today it is part of the Ministry of Science, Technology and Space, responsible for all civilian space research and development. Like most other agencies, it has national security interests at heart. It also emphasizes greater education and development among the country's students, and connections to Israel's burgeoning space industry. Of particular interest are nanosatellites and communication satellites. Ultrasat, an astronomy satellite, is a collaboration between NASA and the Weizmann Institute of Science, Israel Aerospace Industries, Space Systems Loral, California Institute of Technology and the Carnegie Institution for Science. Data gathered by its ultraviolet-sensitive detectors are expected to help astronomers understand what's powering such astrophysical phenomena as supernova explosions and gamma-ray bursts.

While the agency does not have its own human space flight programme, it has collaborated with NASA in the past. Israel's first astronaut, Ilan Ramon, flew aboard the ill-fated shuttle *Columbia* as a payload specialist in 2003.

Korea
In 1989, the Republic of Korea established its Korea Aerospace Research Institute (KARI) as the country's primary agency for aerospace and space science studies. It developed Korea's first space launch vehicle, called Naro-1, and established the NARO space centre in the southern part of the country. KARI does research and development into launch systems and trains personnel in science and support capacities. The agency's first Naro-1 launch was 25 August 2009, but the payload did not reach the proper orbit. A second launch also failed, but the third (on 30 January 2013) was successful and Naro-1 became the first South Korean launch vehicle deliver a payload – the Science and Technology Satellite 2C – to orbit. In 2016, Korea and the US signed a space cooperation agreement that provides for Korea to manufacture its lunar orbiter while the US supplies some payloads and communications support for the mission.

Malaysia
The Malaysian government has long been interested in space exploration as both an economic driver and a way to involve those of the Muslim faith in space travel. Another stated objective of its national agency, ANGKASA, is to inspire young Malaysians to study more maths and science. The country has a fleet of

satellites; the first was Tiung SAT, launched in 2000. It performed meteorology and earth imaging as well as cosmic-ray detections. Three MEASATs were launched to provide television and general communications, and RazakSAT was an Earth-observing satellite. The first Malaysian astronaut, Sheikh Muszapan Shukor, went to the ISS aboard a Soyuz TMA-11 spacecraft in 2007.

New Zealand

New Zealand is very much a newcomer to space launches and has long hosted high-altitude balloon flights. With its first launch to space of an Electron rocket on 25 May 2017, the country has set itself up as a commercial rocket launch centre. The rocket was built by the RocketLab company, which has been providing launch systems and other technologies in hopes of removing barriers to commercial participation in space. The country's Ministry of Business, Innovation and Employment is the agency in charge of New Zealand's space efforts, set up in 2016.

More countries are coming online with formal space agencies and institutes. For example, the Arthur C. Clarke Institute of Modern Technologies in Sri Lanka has been charged with developing a space infrastructure for that country which includes the launch of satellites. In early autumn 2017, the Australian government – which has long supported space efforts – announced plans for a formal Australian space agency. Similar development plans are in the works for the United Arab Emirates, the South African National Space Agency, the Brazilian Space Agency and others. Far beyond the 'romance' of space travel, countries are finding the technological, educational, financial and political benefits confer a prestige in the modern world.

In the main, space exploration as a function of a country's gross national product is smaller than other expenditures such as military spending and social programs (health care, education, etc.). However, the payback for each dollar or rouble or euro spent on space exploration is quite high.

According to a US government report called *The Future of Space Commercialization*, the size of the space economy is large and growing larger each year. It concludes:

In 2015 alone, the global market amounted to $323 billion. Commercial infrastructure and systems accounted for 76 percent of that total, with satellite television the largest subsection at $95 billion. The global space launch market's share of that total came in at $6 billion. It can be hard to disaggregate how space benefits twelve particular national economies, but in 2009 (the last available report), the Federal Aviation Administration (FAA) estimated that commercial space transportation and enabled industries generated $208.3 billion in economic activity in the US alone.

The economic numbers continue to rise as countries and investors see profit in the wild blue yonder. According to a late 2017 investigative piece by the BBC, investments into space exploration are growing rapidly: 'In 2016, the global space economy totalled $329 billion, with three-quarters of that coming from commercial activity – not governments.' Each country that engages in space exploration activities has seen the introduction of products that improve life. Such spinoffs range from solar panels to water-purification systems, improved medical therapies and global communications advances. Spinoffs are nearly ubiquitous, with many so common to our everyday lives that we don't usually stop to think about their connection to space exploration.

The same US report goes on to point out the global aggregate of space exploration efforts may provide an incredibly bright future:

Space is not just about satellite television and global transportation; while not commercial, GPS satellites also underpin personal navigation, such as smartphone GPS use, and timing data used for Internet coordination. Without that data, there could be problems for a range of Internet and cloud-based services ... In the future, emerging space industries may contribute even more the American economy. Space tourism and resource recovery – e.g., mining on planets, moons, and asteroids – in particular may become large parts of that industry. Of course, their viability rests on a range of factors, including costs, future regulation, international problems, and assumptions about technological development. However, there is increasing optimism in these areas of economic production. But the space economy is not just about what happens in orbit, or how that alters life on the ground. The growth of this economy can also contribute to new innovations across all walks of life.

International Cooperation

The idea of international cooperation in space exploration is not a new one, but it is rapidly growing as commercial ventures look toward the new horizons space represents. In 2011, Planetary Society co-founder Dr Louis Friedman made the case for international collaboration in space, pointing out that for years this has been a goal of the major players. He wrote in *The Space Review*:

> I owned an American car built in Australia from an Italian design with Japanese parts. It is trite now to comment on globalisation and interdependency of world industry. The International Space Station is an extraordinary testament to how globalisation and international cooperation have permeated the space program. The United States and Europe have also merged their planning for future robotic space exploration ventures to Mars and the outer planets. If these plans materialize, they will enable much more to be accomplished than could be done in national programs.

Friedman was pointing out there are ways to do what we want to do, if only we can cooperate internationally. Of course, funding for space exploration and science have always been issues in the countries where such money is spent.

International cooperation became the name of the game in the 1970s and has grown, sometimes slowly, since then. The first effort was between the US and Soviet Union with the Apollo–Soyuz mission. Today, the best example is the ISS, with NASA, the ESA countries, Canada, Japan and Russia contributing knowledge, instruments, modules, astronauts and money to the effort.

Most countries seek international participation for several reasons: to get others to share in the expenses of the projects, to gain experts not available in their own agencies, and to share in the prestige of high-profile and successful projects. Politically, a cooperative agreement between countries gives leaders a way to sustain their space programs and other scientific enterprises. It also creates jobs among the partners' workforces, and the spinoffs tend to propagate throughout the world.

International cooperation also brings technological standardisation – what space policy analysts call 'interoperability'. Simply put, if partners are to work together, their technology has to mesh. Such interoperability spreads to industrial partners as well. In addition, language protocols have to be developed. While most spacefarers speak some English (following the tradition of English being the international language of pilots and air traffic control), it has long been required for NASA astronauts to learn Russian. This will remain the case while the US depends on Russia for access to space via the Soyuz craft.

Other projects connect NASA and ESA with a wide variety of global partners. NASA itself has more than 120 global partners, working on oceanographic observations from space, long-term studies of various planets, astronomy missions, information technology systems, aerospace research and sonic boom studies.

ESA is a veritable research village, bringing together many European countries to work together toward common goals in exploration. The agency also works with Russia, ensuring long-term access to space through the spaceport at Kourou in French Guiana. It has signed cooperative agreements with China for future collaborations, and worked with other countries to develop satellites and other infrastructure.

Russian collaboration with India has helped build the Indian space economy and infrastructure through strategic partnership. The Soviet Union was the first to launch Indian satellites for their partner, and is working with the country on the Chandrayaan programme of lunar exploration.

A 'united nations' of scientists worked on many planetary missions flown in the last few decades. For example, the successful Rosetta mission to Comet 67P/Churyumov-Gerasimenko, while mainly an ESA-led mission, also included planetary scientists from the US Mars Express is mainly an ESA mission, but has partnered with Japan, Russia and the US. NASA has supplied the Mars Reconnaissance Orbiter and MAVEN spacecraft as backup data relay systems.

It's not outlandish to think about upcoming human missions to Mars as good opportunities for international cooperation. These will take years to develop and will be quite expensive. Using resources from as many space agencies and countries as possible will help spread the costs out evenly among the participants. When the crews do launch for Mars, carrying international crews trained at NASA or ESA's centres, they will be sent on their way aboard rockets and spacecraft built by NASA, Roscosmos, SpaceX, Blue Origin or the Chinese. The risks will be shared, as will the scientific and technical data generated by the missions. These are the benefits of international cooperation.

One need only look to the example of the ISS to see the benefits of collaborative, cooperative effort. The partners – the US, Russia, Japan, Canada, and ESA – had to work out technological complexities in order to match all the different modules. Each country maintains control of its modules, and countries take turns supplying station commanders for the expeditions. Not only must the astronauts who use the station cooperate, there's a whole infrastructure of ground-based support facilities – from launch support and processing of cargo to mission operations, research and technology development efforts, and of course communications between Earth and the station. Transport to the station is handled by the US and Russia, along with cargo supply ships. All those trips must be coordinated between the partners.

The example of the ISS provides a good blueprint for future cooperation in space. The upcoming 'lunar village' project between China and ESA will be a useful test of the lessons learned from ISS. So will other future missions to the Moon, an asteroid, Mars, and beyond. The biggest lesson that humanity may learn from the Space Age is the need for cooperation in exploration.

The results of that exploration – and the agencies that pursue it – will ripple out across global society. Whole cultures have changed immeasurably thanks to space exploration, and humanity is well on its way to becoming what science fiction writers and dreamers described: a spacefaring society, a civilisation that looks out to the stars while learning to live cooperatively on our home planet as well as in space.

5

Space and Society: Building a Spacefaring Civilisation

Space exploration is a force of nature unto itself that no
other force in society can rival. Not only does that get
people interested in sciences and all the related fields, [but]
it transforms the culture into one that values science and
technology, and that's the culture that innovates.

Neil deGrasse Tyson

Space today remains an ocean that beckons us outward unto
the unknown, but it inevitably turns our vision back toward our
small planet on its shore.

Joe Engle

In order for us to have a future that's exciting and inspiring, it
has to be one where we're a spacefaring civilisation.

Elon Musk

We've been living in the Space Age for well over half a century, but
if we look around the world today, what can we point to that says
'Space Age'? It's more difficult to see when one is immersed in such
a culture and enjoying the benefits without knowing their origins.
However, modern societies have benefitted greatly from what began as
a Cold War race to space. As access to low Earth orbit, the Moon and
Mars grows, our social institutions evolve along with our technology.

The dean of science fiction Robert A. Heinlein attempted to
'look forward' to what our civil societies would be like in the age

of exploration he foresaw. Writing in the 1950s, he looked forward to retiring on the Moon. He also foretold interplanetary travel. He described trips to and from Mars in his stories, as have so many others. He certainly had some intriguing ideas:

> Hospitals for old people on the Moon? Let's not be silly ... or is it silly? Might it not be a logical and necessary outcome of our world today? Space travel we will have, not fifty years from now, but much sooner.

Heinlein might not have pegged the date of future retirement homes on the Moon, but he was right about the space travel. Not quite a decade after he wrote this piece, men were headed to space. Fifty years after that, astronauts from around the world were traveling quite regularly to space. They continue to do so. On Mars, planetary probes do their reconnaissance work so that when humans arrive they'll have the information they need to live and work there. Someday we may see space tourism take off, and maybe there will be colonies on the Moon. Whether they will be populated by retirees or not – well, that's a political and social question as much as a technological one.

Science fiction stories are good for bringing our imaginations to the futures we may want to live in, but they usually pick up in the middle of civilisations immersed in space travel already. Fewer tales start out at a culture's first steps to the cosmos and trace all the developments that led to space flight. Here on Earth, in the twenty-first century, we're still writing our first chapter of the space exploration story, creating and strengthening the institutions that helped create it.

People and ideas help create such a spacefaring society. Money helps ease the way. We didn't just point rockets to space and send people to the Moon overnight, although the development of space flight did occur fairly quickly. As we've seen in the first two chapters of this book, the leap to space came from a confluence of technological, political and financial factors. Along the way, we created other parts of the infrastructure required to build a spacefaring society: educational institutions, business interests and legal and political frameworks.

International Treaties, Space Law and Space Policy

The Space Age would not be complete without laws and policies to guide decision makers as we expand our presence beyond Earth. Some of the laws we follow today come from some very practical questions. For example, who owns the Moon? Anybody? Who is legally allowed to go there? What happens if a launch is aborted and the space vehicle lands in another country? If a mission fails on orbit, what are the laws about rescue or salvage of the damaged equipment? These and many other questions and situations require an agreed-upon set of laws and policies that all countries can follow.

Just as there are laws to govern traffic, commerce, communications and safety here on Earth, so too we have both national and international laws that govern humanity's access to and use of space. Essentially, space laws and treaties set up rules about how we (individuals, companies and countries) behave in space. Political decisions about space flight, space exploration and space utilisation are part of space policy. Space law, treaties and space policy all intersect.

The first glimmerings of what became several treaties about uses of outer space came about when the United States and Soviet Union laid the groundwork for the issues about the use of space. Ultimately, the United Nations stepped in and created the Committee on the Peaceful Uses of Outer Space (COPUOS), with subcommittees focused on scientific, technical and legal aspects of space exploration. Some of its treaties were adopted readily; others still await ratification.

The COPUOS, through its Scientific and Technical and Legal Subcommittees, requires all members from each of the world's member states to agree on terminology, rulings and treaty language. It's often difficult to come to consensus, and the agency continues to grapple with a comprehensive space agreement for the twenty-first century. So far, the UN committee has forged agreements and resolutions that cover most situations. They are:

The Treaty on Principles Governing the Activities of States in the Exploration and Use of Outer Space, including the Moon and Other Celestial Bodies (the 'Outer Space Treaty') (1967, adopted by 104 countries)

The Agreement on the Rescue of Astronauts, the Return of Astronauts and the Return of Objects Launched into Outer Space (the 'Rescue Agreement') (1968)

The Convention on International Liability for Damage Caused by Space Objects (the 'Liability Convention') (1972)

The Convention on Registration of Objects Launched into Outer Space (the 'Registration Convention') (1975)

The Agreement Governing the Activities of States on the Moon and Other Celestial Bodies (the 'Moon Treaty') (1979, with seventeen signatories and not widely ratified)

The Principles Governing the Use by States of Artificial Earth Satellites for International Direct Television Broadcasting (1982)

Essentially these treaties, plus the laws and non-binding resolutions that flow from them, have declared outer space a common domain theoretically accessible to all people on Earth. They spell out the intention that no one country control access to or ownership of space, or other objects in the solar system (particularly the Moon). They also elaborate on what liabilities a space agency has if its equipment causes harm or damage to space stations, capsules or other objects belonging to another country's agency. Anything launched into space remains the property of the nation that launched it, and if anything from a spacecraft lands on Earth it should be returned to its country of origin.

Exploiting Space

The rise of private industry access for space commercialisation – such activities as mining asteroids, for example, and the creation of commercial spaceports and resorts – influences new and expanded regulations. Compliance with international law is balanced against the abilities of countries and companies to get investors interested in these commercial activities. Nations now gearing up (or those with plans) for space exploration want to ensure their own access and that existing spacefaring nations will not unfairly monopolise available resources.

One possible way to deal with these future activities would be for the United Nations to extend its Convention on the Law of the Sea to extend to space as well. This established rules and regulations about how countries use the ocean resources available within their own borders as well as the open sea. Extending it to space would have the effect of placing similar rules on space resources. Some countries are considering this as they seek to draft their own laws about exploiting resources in space.

One of the most contentious and serious domains for space law is the exploitation of natural resources. There could be great rewards for asteroid miners or lunar resource traders, for example, but who owns these objects, or the rights to their resources? The Moon treaty says that, like Antarctica, the Moon belongs to all countries; no one country can seize it as territory. It should be treated more as a scientific 'lab', where exploration and experimentation can take place. That doesn't mean China or the US can't land there and immediately mine for water or materials to help build bases, but it does mean the two countries can't strike a deal to 'own' the Moon. The use of natural resources in space is likely to be a hot topic in the very near future, and countries are developing laws governing access to such resources.

The US, for example, has taken steps to open up resource extraction to American businesses and individuals. On 26 November 2015, President Obama signed the Commercial Space Launch Competitiveness Act. Under its provisions, US citizens and commercial interests are given the explicit right to claim and own resources available from space objects. This includes water ice and minerals found on asteroids. These rights are recognised under international law and treaties. The idea is to use the law to encourage the mining of an asteroid or other resource in the course of developing a mature space economy and the infrastructure needed to utilise the resources.

The US isn't the only country to work up secure legal frameworks for utilising space resources. The tiny European country of Luxembourg, long known as a business-friendly environment, gave substantial financial support to its own satellite company, SES, making it the second-largest satellite operator. Based on that experience, the country has taken a deep interest in asteroid mining. In 2016, the country's Ministry of Economy announced its Space Resources Initiative, which developed legal and regulatory frameworks regarding the future ownership of minerals and other materials extracted from asteroids. Along with a pledge of financial support from companies wanting to develop space mining, Luxembourg attracted major players from the US.

Communications and Sensing Satellites

The Space Age brought satellites for communications, sensing and, of course, reconnaissance. The language of the UN treaties suggests such space communications (unless for national security or proprietary reasons) should promote mutual cultural and scientific exchanges of information and assist in educational, social and economic development.

Over the years, the treaty has helped countries develop rules about where in space these communication and sensing satellites may be placed. Geostationary satellites (those that orbit at the same rate Earth turns, and thus stay 'stationary' over a given point on the planet) are required to occupy a single ring above the equator, approximately 35,800 kilometres from Earth. Only a limited number of satellites can occupy this region due to spacing requirements between 'birds'. This has led to disagreements between countries arguing over the same orbital berth. Such claims are arbitrated through the International Telecommunications Union, which is the UN agency charged with control of orbital slots.

Nuclear Safeguards

One of the biggest concerns dealt with in space treaties is the use of nuclear power sources on spacecraft. In the early years of the Space Age, the US and Soviet militaries each suspected the other of working to orbit nuclear weapons. Today, the agreements regarding nuclear power sources (written in 1992), are more concerned with the safety of such equipment as radio-thermal isotope generators (such as those used aboard various planetary missions). It states:

States launching space objects with nuclear power sources on board shall endeavour to protect individuals, populations and the biosphere against radiological hazards. The design and use of space objects with nuclear power sources on board shall ensure, with a high degree of confidence, that the hazards, in foreseeable operational or accidental circumstances, are kept below acceptable levels.

The concerns are understandable; if an RTG-equipped spacecraft were to fall back to Earth in a launch accident, for example, there's a concern the capsule containing the radioactive power source could contaminate the area where it falls. The chances of such an accident are small, but an agreement covering such a possibility gives guidance about what to do if it does happen.

Regulating the Space Station

The creation of the International Space Station prompted the UN to develop a treaty for those involved in its construction and use. It formalised the arrangement between Canada, ESA, Japan, Russia and the US. As mentioned in chapter 4, NASA takes the lead on the station, but the treaty allows each nation to have control over the module(s) it contributed to the station. So, for example, the Japanese have control and jurisdiction over the Kibo module, which is a large 'container' of science labs.

Space Law

Although there are UN treaties and agreements governing global access to space, individual countries have passed thousands of laws regarding their own space activities and industries, as required by the Outer Space Treaty. Some of these laws regulate the countries' participation in international cooperative efforts. For example, in 1962 the US wrote and passed the Communications Satellite Act, regulating the commercialisation of space communications. It was a response to the tremendous technological growth in communications at the start of the Space Race. The Act provided for global participation in the establishment and use of communications satellites programs, and authorised a publicly held corporation called Comsat to operate one. It was part of Intelsat, an intergovernmental organisation coordinating communication satellite deployment and use around the world.

Generally, US space laws govern spacecraft operations by governmental, private and launch servicing organisations, and draws upon laws in other areas, including telecommunication and remote sensing. Although there are regulations governing human space flight, the advent of space tourism (for example) may involve the application of flight regulations already in use for commercial airports and aircraft use. Laws involving technology transfer and liability from space debris will need to be rewritten or extended to include space activities not already covered. In the US, space laws originate in Congress (with input from government agencies and affected citizens and businesses), eventually are passed by the Senate, and then are sent to the President for signature. Then, it's up to the related agencies (NASA, FAA, etc.) to implement the rules and regulations enabled by the law.

Russian space law is aimed largely at making sure the country has continued access to space, and ensuring defensive capabilities while maintaining control over how international treaties affect its armaments and armed forces. Like most other countries, Russia also continues to study Earth from space, and uses its access to space (currently to the ISS) to develop and conduct science experiments and related technologies. Russian space activity is supervised by the Council of Ministers, and the President is responsible for implementation. Roscosmos is responsible for carrying out exploration activities.

China – a relative newcomer to space exploration – is in the process of writing its own set of space laws. Given that its space activities are largely controlled by the country's military establishment, China's efforts are aimed at policy and regulatory frameworks fitting into international agreements already in place.

Japan, too, has written laws to work within the frameworks of the international treaties to which it is a signatory. While the country has been launching rockets since 1970, and is the only Asian country participating in the ISS (with its Kibo laboratory), its laws do not allow other than peaceful uses of space. The country has joined twenty other nations that have passed laws expanding business activities to space. In November 2016 the country published its Space Activities Act, which establishes a licensing system for rocket launches and satellite operations by private companies.

For much of its history, India's legal policies about space were primarily guidelines for satellite communications and remote sensing. As the Indian Space Research Organisation (ISRO) has moved into satellite launches for other countries, the country has drafted legislation that unites civilian activities with strategic and military components. As the country's political establishment debates these proposed laws, ISRO (which is a civilian agency) and others have raised questions and cited their strictly scientific aims for space research. This has led to debates about space policy within the Indian scientific and manufacturing communities. In a 2016 Space Security Initiative brief from the University of Washington School of International Studies, journalist Deep Pal wrote:

> A declared space law will help India defend its interests and strategic goals for space exploration and use. As more countries elevate their spacefaring capabilities, tomorrow's rivalries will be about the weaponisation of space, protecting assets in orbit, handling space debris, and regulating spectrum allocation [...] Declaring its stand now and making the broad outlines of its objectives public through a well-articulated space law will ensure that India is not left sitting outside such a global regime as it had to do on the issue of nonproliferation.

The Cost of Going to Space

Space missions can be quite expensive. Questions of just how costly are often raised when government budgets are set. Since taxpayers are asked to foot the bill for every government expenditure, whether for military defence, salaries of government workers, social programs, educational programs and, yes, space exploration, people are rightly concerned about the costs and benefits of space exploration.

It costs a lot to go to space. For example, the US shuttle programme cost about $196 billion, averaging about $450 million per mission. The Hubble Space Telescope cost $1.5 billion to build, while ESA's Herschel Space Observatory was estimated to be about €1.1 billion. India's Mars Orbiter Mission cost the equivalent of $73 million just to get it to the Red Planet. When the US wants to send an astronaut to space aboard the Russian Soyuz capsule, each seat costs about $75 million. The cost of the Juno mission spacecraft was $1.1 billion. None of these costs include ongoing expenses such as salaries and mission operations. As you can see, space exploration has some very real and serious costs attached to it, expenses the countries involved are willing to pay for the prestige and scientific returns.

In the US, it has been variously estimated that every dollar Americans spend on NASA returns at least ten dollars to the country's economy – much of that in commercial goods and services. As an employer/contracting agency, NASA's earliest budgets supported less than 10,000 employees; that figure is now over 18,000 employees, plus perhaps three times that many working for NASA contractors.

Space Agency Budgets

To give an idea of the costs of space exploration, for the top six countries involved in space exploration the budget numbers were in the billions, for 2016:

NASA	$19.3bn
Roscosmos (Russia)	140bn roubles ($2.3bn)
China	13bn yuan ($1.9bn)
JAXA (Japan)	154.1bn yen ($1.3bn)
ESA (Europe)	5.25bn Euros ($6.1bn)
ISRO (India)	77bn rupees ($1.2bn)

Other countries are spending proportionally less. A more nuanced analysis takes into account costs as a portion of a country's GNP; however, for our purposes these figures do give a good idea of the money being spent on space exploration. The countries involved see it as an investment in their futures – not just in exploration but in increased productivity, access to technology, better education and, of course, highly paid jobs available to their citizens. The spinoffs from space exploration fit across a spectrum of technological, financial, medical, communications, and other industries.

What Does It Cost Me?

Often the question is asked by voters, 'What does space cost me?' For Americans, the cost per taxpayer is a frequent measure of the worth of a programme. The answer in terms of cold hard cash is fairly easy to figure out. NASA's part of what a person pays in taxes is a pretty small piece of the pie. For every federal tax dollar a citizen pays, about half a cent goes to NASA. If an average tax bill is $5,000, for example, that comes to about $25.00 per year. That's a pretty small investment for something with good returns to the economy.

Education Is the Key to Space

In most countries around the world today, science, technology, engineering and mathematics (STEM) subjects are considered important for people working in space exploration. Such education starts well before students get to college or other higher education, or join the military.

During the Space Race, it was clear that more science and technical education was needed as the countries ramped up to meet the personnel requirements of a space infrastructure that was being built almost overnight. In the US, the government passed the National Defense Education Act (NDEA) in 1958. Not only was it a response to Sputnik, but it became very clear the nation – which already had a very good educational system – needed to train even more scientists and technical experts than it already had. So, money flowed to school districts for curriculum upgrades and science teachers, to colleges for advanced courses of study. The National Science Foundation (NSF) was granted a half billion dollars over twenty years for maths and science curriculum upgrades and teacher development.

The state of science education today is a much-debated topic, particularly in the US. There may not be the 'Sputnik' impetus anymore, but the need for solid maths and science instruction remains important. In addition to classroom work, students can get involved in hands-on science experiments. For example, CubeSats are miniaturised satellites that can carry small instrument packages to space. In 2013, students at Thomas Jefferson High School in Virginia worked on the TJ3Sat, with help from an aerospace company. It was the first-ever orbiting satellite to be built by high school students. It was launched by NASA and stayed in orbit until 2015. Another student-built project, called STMSAT-1, was launched from the ISS in 2016. Students collaborated with NASA's Goddard Space Flight Center on the project, which was designed to take images of Earth's surface.

In the UK and the Netherlands, a group of amateur radio enthusiasts collaborated with educators to create FUNCube-1 to provide real-time data for classroom applications and lessons. Students and amateur 'hams' around the world tune in to its transmissions.

In the US, STEM efforts also aim at bringing diverse populations into science. While steps are being taken to include interested women and girls into the scientific realm, there are still many societal and cultural barriers to overcome. Women make up a smaller part of the scientific workforce, and face discrimination at all levels of their scientific careers. The same is true for students of colour, often overlooked for science careers or outright discouraged from pursuing them.

Academic Steps to Space

Education and academic research are an integral part of space exploration. With robust universities and research institutes in place around the world, it was only natural that early rocket experimentation first took place in academic labs. Robert Goddard's work with the first liquid-fuelled rockets was done while he was on the faculty of Clark University physics department. Wernher von Braun's work with Herman Oberth at the Technical University of Berlin sowed the seeds for the later development of the V-2 rocket. In the early years of the Soviet Union, Konstantin Tsiolkovsky pursued academic research into the theories of rocketry, which provided much-needed expertise for government-built labs and design bureaus for its own missile programs.

During the headlong rush of the Space Race, the US looked for research talent in universities, colleges and the military. NASA had a growing collection of labs and institutes. Commercial space interests (aerospace companies) were also involved, often working jointly with NASA and academic institutions in large teams. Universities have also been the academic training grounds for space crews, supplying expertise in physics, mathematics, engineering, propulsion systems, and other subjects required for their missions. Today, many space explorers have degrees in astrophysics, engineering, geology, medicine – all vital topics.

There are hundreds of universities and colleges with academic departments focused on space-related topics around the world. In the US, for example, in subjects such as aerospace engineering and space science, sixty-eight countries have colleges or universities offering degrees in the subject. Here is a sampling of the top-ranked institutions for the major spacefaring nations as of 2017.

China

Beijing Institute of Technology; Beihang University School of Aeronautic Science and Engineering; Harbin Institute of Technology; Northwestern Polytechnic University School of Aeronautics; Nanjing University of Aeronautics and Astronautics; Peking University; Tsinghua University School of Aerospace; University of Science and Technology, School of Earth and Space Sciences; Xiamen University (Physics, Mechanical and Electrical Engineering); Zhejiang University School of Aeronautics and Astronautics

European Union

International Space University; Heidelberg University; Leiden University; Munich Technical University; Pierre and Marie Curie University; University of Munich; University of Padua

Japan

Nagoya University; Kyoto University; Kyushu Institute of Technology; Osaka University; Nagoya University

Russian Federation

Bauman Moscow State Technical University; Moscow Aviation Institute; Moscow Institute of Physics and Technology; M.V. Lomonosov Moscow State University; Samara State Aerospace University; Siberian State Aerospace University

United States

Georgia Institute of Technology; Massachusetts Institute of Technology; California Institute of Technology; Purdue University;

Stanford University; Texas A&M University; University of Arizona; University of California; University of Colorado at Boulder; University of Illinois Urbana-Champaign; University of Maryland, College Park; University of Michigan; University of Texas at Austin

United Kingdom
Bristol University; Imperial College London; Loughborough University; University of Bath; University of Cambridge; University of Southampton; University of Surrey; University of Edinburgh

At most institutions that offer degrees in space-related technologies, opportunities exist for students at both the graduate and undergraduate level to participate in missions. In Germany, the Max Planck Institute for Solar System Research led a successful team effort to explore Comet 67P/Churyumov-Gerasimenko using ESA's Rosetta spacecraft. Team members included planetary scientists, graduate students, interns and post-doctoral researchers.

These days, most missions involve students at some level. The University of Colorado's Laboratory for Atmospheric and Space Physics was the first to allow undergraduate students to routinely work in spacecraft control centres, monitoring and commanding planetary and earth-observing missions. Its graduate students work in engineering, mission data analysis, spacecraft system design, testing and construction of instrumentation.

At MIT's Kavli Institute, students are involved intimately in research ranging from astrophysics to technology development. The US Air Force Institute of Technology offers military personnel hands-on training with aspects of aeronautics and space science technology.

Students at some universities have formed their own space societies, or joined more mainstream groups such as Students for the Exploration and Development of Space (SEDS). Cambridge University Spaceflight is a group that regularly involves graduate and undergraduates in building and testing instruments to fly aboard high-altitude balloons and rockets. Members work with the physics and engineering departments during outreach events to share their experiences.

Education in Space Policy and Law

There is a growing body of case law involved with space exploration, ranging from access to space to control and jurisdiction of space assets, intellectual property and other legal topics. A number of universities offer coursework and practical experience in the topics. The field is still quite new, but already legal minds have been tackling topics and rulings using extensions of existing aviation law (for example) or working with governments and agencies to formulate new laws. Commercial aspects alone would constitute a whole legal sub-specialty. For example, increased access to space will someday result in a rapid growth of space settlements. Existing laws and treaties do not always cover those situations. Nor are all the laws on the books applicable to space tourism. There will come a day when commercial transport to space will be as viable as passenger flights in the early days of aviation. When that happens, questions of liability and responsibility will become part of the contracts of carriage between the space launch companies and their clients.

These and many other situations require a new cadre of legal experts – space lawyers, if you will. They will join the ranks of policy makers and advise political and institutional leaders about the 'rules' of space exploration. So, where will they learn their trade?

The European Space Agency routinely offers summer courses on space law and policy. In France, the University of Paris-Sud has a master's programme that emphasises legal aspects of space exploration, and supplies commercial space companies with its graduates. In the Netherlands, the Universiteit Leiden operates an international institute focused on space law and policy issues regarding aviation and space activities.

In the US, the University of Nebraska offers programs up to and including doctoral work on space and cyber law. The University of Mississippi's law school offers programs that deal with developing case law to cover remote sensing issues. These are in addition to institutes in Washington, D.C. squarely focused on issues of space law.

Public Interest in Space Exploration

In the early years of the Space Age, people were fascinated with astronauts, their missions, and the undeniable sense of a 'race' between two major world powers. After the Moon landings, press attention didn't linger long on space shots and missions unless something happened to attract it. As usual, spectacular images will grab some attention, and disasters are always good for a few headlines.

However, public interest in space seems to remain. In a way, that's good – it indicates space exploration is on a safe footing. On the other hand, if space exploration activities gather little attention, people (or their representatives) may not be eager to continue funding them. To share stories of their explorations, discoveries, and achievements, the world's space agencies maintain public outreach programs, sharing their work with taxpayers through webpages, museum exhibits and other educational programs. It's to their – and our – benefit to keep attention focused on humanity's future in space. Not only does it expand our interests, but the effects of expanded exploration are felt around the world.

Space and the Arts

Society embraces space exploration themes in many other ways, particularly through the arts and entertainment industries. Music, particularly rock, pop, and progressive genres adapted space themes among group names and actual tunes. The rock-and-roll group The Astronauts toured the US in the 1960s, trading on the popular fascination with people traveling to space, while playing surfer music. After the launch of the Telstar communications satellite in 1962, the British rock group The Tornados debuted a hit instrumental called 'Telstar'. In 1970, the group Jefferson Starship devoted an entire concept album to the idea of space travel in their landmark work *Blows Against the Empire*. The Alan Parsons Project recognised the Voyager mission to the planets in one of the pieces on their *Pyramid* album.

Around the same time (1970s), synthesizer-based music was becoming more popular. These electronic instruments provided otherworldly sounds, allowing composers to develop a new genre called 'space music'. Composing music for planetarium show soundtracks since the 1970s, the artist Geodesium helped define this style of ethereal ambient music, scoring soundtracks to send audiences floating through nebulae, past supernova explosions, across galaxies and into black holes.

The rise of the Space Age brought new inspiration for visual artists. 'Space art', sometimes called 'astronomical art', was around long before the Space Race, most notably gracing the covers of pulp fiction magazines and science fiction books. One of the first artists of the genre, Chesley Bonestell, used his paintings to popularise space travel. His work ranged from otherworldly landscapes to nearly photorealistic scenes of spacecraft in orbit around Earth and other words. Other artists followed in Bonestell's tradition, creating visions of distant worlds.

As planetary probes returned real-life images of outer space, the work of space artists incorporated them. Much space art also depicts future missions, such as Pamela Lee's visions of explorers on the Moon and Mars. Aerospace companies also work with space artists on animations and graphics to illustrate future missions and design concepts, and NASA itself has a history of commissioning space art to illustrate its reports. Today, space artists – among them Lynette Cook, Don Davis, David A. Hardy, Lee and Lucy West-Binnall – work directly with scientists and film producers to create views of the cosmos.

More than 120 artists who specialise in space themes have joined the International Association of Astronomical Artists (IAAA), among them several former astronauts. Whether their works adorn someone's wall, are reprinted on a magazine cover, projected onto a planetarium dome or used in a movie, the visions of space artists reveal new views of the cosmos to the public.

Popular Culture

Of course, many people can readily name space exploration-themed movies: *Star Trek*, *Star Wars*, *2001: A Space Odyssey*, *Interstellar*, *Gravity*, *Close Encounters of the Third Kind* and many, many others. As mentioned throughout this book, science fiction has been telling stories of the future in space for more than a century.

Some books, such as Kim Stanley Robinson's *Mars Trilogy* and Greg Bear's *Mars Crossing*, focus on a near future on Mars. Others, such as Asimov's *Foundation* series, tell stories of humans in far future grappling with galaxy-wide problems.

Real space images often find their way into other art forms; *Star Trek*, for example, often used images from the Hubble Space Telescope as set decoration.

Space themes are a big part of video games and gaming apps. The first digital games, such as Asteroids, were little more than 'shooting' galleries, where users would try to take out targets. Today, the gaming industry has many titles where the action takes place in space. There are also educational games such as NASA's *Space Place*. In it, users can travel through the galaxy, use satellites and study comets. One of the most popular space simulators is *Kerbal Space Program*, which lets players create their own astronauts, vehicles and science experiments.

It's also not unusual to see space-related ideas in advertising, fashion events (the Chanel 2017 fall/winter show featured a catwalk decorated with rockets), marketing backdrops and other public places. It would seem that space travel gives a specific impression that help companies sell ideas and products.

Virtual Space

Virtual reality (VR) headsets provide a new playground for enthusiasts to experience space exploration while never leaving Earth. Virtual reality has long been used for training purposes – astronauts take virtual spacewalks, learn to use new equipment and simulate missions long before they ever take a step into space. When the Hubble Space Telescope needed repairs and refurbishments, astronauts used VR headsets in virtual environments to learn the tricks of the trade. Today, NASA's VR training simulations include full tours of the ISS and other vehicles, and the agency has turned it into a consumer product available on the Oculus Rift system.

Space simulations are a great opportunity for the fledgling VR industry, seeking ever more immersive experiences for users. Similar to how a planetarium immerses an audience with content, such packages as *Mission: ISS* (the Oculus Rift project mentioned above) aim to give users the same chance to explore space as astronauts have, but from the comfort of their home.

Augmented reality (AR) units such as Microsoft's Hololens allow users to project images over real-life scenes. Developers see it being useful for educational outreach efforts, as students 'train' to go to Mars or explore under the depths of Earth's oceans.

Planetarium Facilities, Science Centres and Space Camp

Planetarium facilities and science centres around the world are part and parcel of the cultural interest in astronomy and space science. The term 'planetarium' can refer to the star projector commonly installed in a domed theatre, although the entire building is sometimes referred to as a planetarium. Most people know of them as places to learn about space and astronomy, which makes them a go-to facility whenever there's a space-based event.

The modern age of the planetarium began in Germany with the building of the first instrument by the company Carl Zeiss, installed in the Zeiss-Planetarium in Jena. That was in 1923. In the following decades, major planetarium facilities were built in many large cities, including Chicago, Los Angeles, Moscow, Munich and New York. In the years between 1923 and 1957, a total of seventy-six planetarium facilities were built around the world.

The aftermath of the Second World War split Carl Zeiss into two parts (one in East Germany and one in West Germany), and both sides had to rebuild in the aftermath. That left few choices for planetarium instruments. In Japan, the Minolta company began making planetarium star projectors in 1950, joined by GOTO Inc. in 1959. In the US, writer and inventor Armand Spitz built his own star projector company and sold hundreds of his Spitz instruments across the country. With the rise of digital fulldome projection systems beginning in the late 1990s, companies such as Digitalis Education Solutions, Evans & Sutherland, Science First, Sky-Skan, Inc. (all in the US), Emerald Digital Systems (Israel), RSA Cosmos (France), and SCISS (Sweden) have stepped into the market for digital video systems that provide full-dome immersive experiences.

A community of content producers, including those within facilities as well as freelance filmmakers, serve the need for fulldome videos, while in-house lecturers and presenters interact with the public for live performances. The present author is one such producer, having participated in the creation of more than two dozen fulldome films for distribution around the world. Fulldome shows range from astronomy and space exploration documentaries to films presenting other sciences (oceanography, for example), mathematics, history and entertainment.

In addition to fulldome videos, many planetarium operators give live 'star talks', where they present the objects visible in the night sky. Teachers and lecturers use the planetarium to teach hands-on astronomy lessons on lunar phases, orbital motions and other concepts.

The growth in the number of planetarium facilities mirrors the rise and fall of the Space Race, particularly in school/district and college/university facilities in the US. Nearly 600 were installed in the years 1957–1969, with a peak number of eighty-nine built in 1969 alone. Since that time, the rate of construction of new facilities has fallen rather dramatically and fewer are built each year.

Early in the Space Race, the Morehead Planetarium in North Carolina was built to teach astronomy to students and the public, but was also pressed into service as an astronomy training facility for NASA. From 1959 to 1975, Mercury, Gemini, Apollo, and Skylab astronauts were taught the basics of celestial navigation at Morehead. Twenty-six Apollo astronauts learned star identification and celestial navigation at the Griffith Observatory and Planetarium in Los Angeles. In Russia, the Moscow Planetarium was used by cosmonauts for similar work.

With more experience in space flight and improved navigational systems the need for such hands-on training went away, but planetarium facilities have continued to be an important way to teach astronomy to students and the public. There are currently more than 4,000 such facilities in the world; about two-thirds are in schools and universities, and the rest are in science centres and museums. A few are also co-located with observatories, presenting a full range of astronomy experiences for visitors.

There are about 400 science centres in the world, ranging from site-specific facilities at NASA Kennedy Space Flight Center to such institutions as the American Museum of Natural History (New York City), the China Science and Technology Museum (Beijing), the Deutsches Museum (Munich), the National Science Centre (Kuala Lumpur), the National Space Centre (Leicester), the Science Museum (London), Scienceworks (Melbourne), Tsukuba Space Centre (Japan) and many others. Taken together, planetariums and science centres are a unique part of science education's 'informal' outreach efforts and engage the public's continuing interest in astronomy and space science.

For a more hands-on approach, hundreds of thousands of people from around the world have attended Space Camp, where both children and adults experience a simulated space mission. The idea was first suggested by Wernher von Braun in 1977, when he saw children visiting the Alabama Space and Rocket Center. He wondered why kids couldn't have a 'science camp' just as they had music and summer camps. Space Camp opened in 1982 in Huntsville, Alabama, and offered attendees training classes, simulators and tours of the nearby Marshall Space Flight Center. Over the years Space Camp has hosted children and adults from the US and sixty other countries, and counts among its alumni astronauts from both NASA and ESA.

Space Societies

Special interest groups and space societies – groups organised around specific aspects of space exploration – are a result of the rise of rocket technology, and blossomed during the Space Race. The oldest groups to get started were the American Interplanetary Society (1930), the British Interplanetary Society (1933) and the German Verein für Raumschiffahrt (1927). The AIS was founded by science fiction writers and began publishing a research journal. It eventually merged with the Institute of Aerospace Sciences to become the well-known American Institute of Aeronautics and Astronautics (AIAA), which has 30,000 members and a network of student branches as well. The BIS counted among its early members such luminaries as Sir Arthur C. Clarke and Sir Patrick Moore, and continues its advocacy for space exploration into the twenty-first century.

Today, there are dozens of space advocacy organisations whose members spend time and effort to further the cause of space exploration. Many have annual meetings and publish substantial papers and treatises on aspects of space exploration. Below are some of the best-known, many based in the US, and a number with global outreach:

Alliance for Space Development A space policy interest group that gathers likeminded groups to help foster the development and settlement of space. It was founded by executives of two other groups, the National Space Society and the Space Frontier Foundation.

American Astronautical Society Founded in 1954, bringing together space scientists, technical experts and other people interested in shaping the US space program's goals and projects.

British Interplanetary Society Founded in 1933, this group (which counted Arthur C. Clarke as one of its early members) promotes the exploration of new concepts for space travel and exploration.

Canadian Space Society A group formed to encourage international cooperation and collaboration in space exploration.

International Astronautical Federation Founded in 1941 to connect scientists involved in space exploration. This space advocacy body has more than 300 members working at space agencies and companies in sixty-six countries.

Mars Society Formed to advocate for the exploration and settlement of Mars through public outreach, and provide support for both private and government-funded Mars missions.

National Space Society This group is an independent educational space advocacy community dedicated to the formation of a spacefaring society. It was originally the National Space Institute and merged with the L5 Society in 1987. There are more than fifty chapters around the world.

Planetary Society A worldwide organisation of people involved in public research and political support for space exploration and astronomy. It was founded by Carl Sagan, Louis Friedman and Frank Murray, and has gone from being a planetary exploration advocate to participating in solar sail experiments with NASA.

SETI Institute This group's mission is to explore, understand, and explain the origin of nature and life in the universe. It was founded in 1984, and as a research institute employs more than 130 scientists in grant- and donation-funded research. SETI stands for the 'Search for ExtraTerrestrial Intelligence', and the group maintains the Allen Telescope Array of radio observatories that search for signals from possible alien life.

Space Frontier Foundation Seeks to transform the space industry through the power of free enterprise. Made up of space activists, engineers, scientists, media and entrepreneurs who meet regularly to advance the concept of 'NewSpace' – a new way of doing business in space exploration.

Space Foundation   An advocacy group of long standing that seeks to advance all sectors of space exploration. It has members among the world's aerospace and science communities.

Students for the Exploration and Development of Space   A student-led organisation that brings young people directly into contact with space exploration projects as part of their career preparation.

Anyone with an interest in any aspect of space travel, astronomy or settlements in space can join these groups and participate in meetings and discussions about the future.

These organisations often advocate for specific types of missions, or the advancement of space policy. For example, the Mars Society's main interest is in pushing the Mars Direct mission, and it has been involved in studies at Mars analogue stations in the US and Arctic.

The Mars Society is an outgrowth of a series of meetings in the 1980s that were held to plan future missions to the Red Planet. There are chapters in the US and Europe. Although the group was first formed from graduate and undergraduate students at the University of Colorado, it has now spread beyond its academic roots and has adherents in space agencies, universities, research institutes and the general public.

Societies such as these have always been one of the best ways to gather like-minded enthusiasts (and space exploration professionals) together to plan for a future in space.

Space Exploration Technologies and Their Benefits

A civilisation doesn't go to space without creating whole new technologies which eventually find their way into the private sector. As we've seen, the development of rockets for war has led to men on the Moon and probes on other planets. Telecommunications satellites first built to relay our thoughts and words around the globe are now space utilities, carrying entertainment programs, communications, imagery and more.

Technologies that allow astronauts to live comfortably in orbit are improving our homes and work places. Space exploration continues to provide employment scientists, technical experts, lawyers and space policymakers, educators, writers, artists, and medical personnel. Thousands of companies are involved in building everything from space capsules to space suits and creating space food and life support sensors. It would be impossible to list every product inspired by space – and indeed, NASA has an entire web site devoted to such technologies – but it's worth looking here at some good examples.

Transferring Technology from Space to Your Home
The path that space technology takes from space agency to the general public is called 'technology transfer'. NASA's charter requires it to provide a formal programme that enables private industry and other entities to license space-related technologies for further development and sale to the public (provided they are not subject to security restrictions). The idea is that taxpayer-funded discoveries, patents and technologies should be transferred to the private sector for use in publicly available products. Each NASA centre maintains an office of technology transfer.

The transfer of space-related technology isn't limited to NASA. ESA maintains a similar programme, providing information about its available 'products' (software, apps, hardware innovations, patent licences) to entrepreneurs via websites and publications. There are many ways that ideas and products from space agencies make their way into our daily lives.

In 2016, three students at Cranfield University (UK), used high-resolution imaging technology to create an aerial vehicle to track wildlife in Africa with the hope of catching poachers before they kill endangered species. NASA Johnson Space Center developed a real-time locating system for use in space that can also be used by emergency and military agencies to track personnel. The scratch-resistant lenses of your glasses and the ear thermometer your doctor uses to assess your temperature are based on NASA technology. The glittery gold-coloured space blanket that spacecraft designers use to protect critical technology now shows up in first-aid kits, survival kits and camping gear stores around the world.

Self-driving John Deere tractors benefit from a long-time partnership with NASA Jet Propulsion Laboratory, which developed a tool that streamed satellite tracking data via the internet to the tractor's navigation system. Our phones and cameras use digital image sensors first developed for planetary missions. CMOS (complementary metal-oxide semiconductor) sensors are everywhere, taking in our personal moments the same way they take in distant worlds for science. Many people own beds that use a memory foam material originally developed by NASA for cockpit seats. Air purification technology first developed for long-term missions (such as the ISS campaigns) is finding its way into scrubbers used in homes, hospitals and other places where clean indoor air is important.

The world's satellites constantly monitor weather systems and study seasonal changes throughout the year. Among other things, these multi-spectral systems can monitor drought conditions, changes in crops and other vegetation, and follow natural disasters in near-real time. Most of us understand the need for accurate weather and climate information all too well, particularly during stormy times such as monsoon, hurricane, and typhoon seasons. Data from these satellites often help scientists identify possible drought conditions, the movements of insects such as the pine beetle, and the change of sea ice at the poles.

As the world moves toward more use of alternative energies, solar cells are providing more electrical power. These are based on space technologies used to power orbiting spacecraft such as the ISS. Recent developments in the solar energy front include the development of flexible solar cells that can be made at a fraction of the cost of rigid cell technology. Not only do these find their way into space technology, but they are becoming available for off-the-grid energy packages as home and business owners seek out alternative energy sources.

Once the exclusive province of militaries, GPS, GLONASS and the Galileo global navigation satellite systems are now used in many civilian applications, including international banking, ship and train navigation, and the delivery services that bring packages to our homes and businesses.

Looking Outward ... and Back

This chapter provides just a taste of what it means to be part of a spacefaring civilisation. Many sectors of society – from education and law to entertainment and business – are affected or influenced by space exploration. In chapters 9 and 10 we take a forward look at what the future holds for space exploration. There are human Mars missions being planned, and – for the wealthy – chances to take a trip to space. Early in 2017, two space tourists put down a deposit on a future round-trip to the Moon aboard a SpaceX mission. It's not clear who they are or how much they're paying, but a look at the cost to send an astronaut to the ISS for a trip aboard Soyuz gives a good indication that they are quite wealthy. NASA pays Roscosmos around $81 million for a round-trip ticket for one astronaut. That includes all the training and consumables for the flight. If, in the future, someone wanted to buy a ride on the new Starliner CST-100, it would cost them around $58 million.

Looking further out, future Mars missions could be financed by individuals who want to pay their own way to a new home by spending about what it would cost them to buy a house on Earth. It's an interesting proposition, and if history is any indication there will likely be a good number of people who might spend the money to be the first human Martians.

It has only been a generation or two since we first sent men to space. One of the most important lessons we've learned is that our planet is a fragile oasis in space. It's an important realisation and one necessary to our evolution as a spacefaring civilisation. We're still learning to protect the planet our space probes have shown us from ever-greater distances. Mounting missions to other worlds is still very much a work in progress, and will rely on technologies we are only now developing and perfecting.

Still, our attitudes will shift over the next decades, as the costs of space exploration come down. The first generations of space explorers are already passing the torch to the next, figuratively giving the keys to the space station, the Moon and the first colonies on Mars to new generations of astronauts who grew up always seeing missions launching to space. For them, it will be a much more natural step to new worlds. Soon enough, our children – or our children's children – will lead the way to Mars and beyond. Indeed, for the first generations of space explorers – well, it might not be all that long before Robert Heinlein's idea of retirement villages on the Moon comes true.

6

Private Industry: Commercial Steps to Space

I believe that space travel will one day become as common as airline travel is today. I'm convinced, however, that the true future of space travel does not lie with government agencies – NASA is still obsessed with the idea that the primary purpose of the space program is science – but real progress will come from private companies competing to provide the ultimate adventure ride, and NASA will receive the trickle-down benefits.

Buzz Aldrin, *Magnificent Desolation*

When something is important enough, you do it, even if the odds are not in your favor.

Elon Musk, founder of SpaceX

Expect the unexpected, and whenever possible, be the unexpected.

Lynda Barry

In the early days of space travel science fiction stories, writer Robert A. Heinlein spun a tale about a self-made millionaire and wheeler-dealer named Delos D. Harriman. In *The Man Who Sold the Moon*, Harriman planned a private-venture rocket to take him to the Moon. He exploits business, legal and charitable sectors to finagle plans. The rocket eventually takes off, but with a trained pilot in his place. Harriman is a combination of robber baron, audacious dreamer and hard-headed capitalist. He sees the benefits of space exploration and adventure of space travel, all while working to get what he wants. He cadges investment capital from fellow millionaires, who sign on with visions of dollar signs in their heads. He plays investors and countries off one another, and makes big promises about handsome returns. The gamble pays off, not just rewarding Harriman and his partners with fabulous profits but bringing the gift of interplanetary travel to the world. Ultimately his company builds a lunar colony, runs a profitable space transport business and changes the face of exploration. Finally Harriman makes his way to the Moon, where he dies a much older and very happy man. His is a classic story of dreaming and achieving big, which resonated strongly in post-war America.

Heinlein created Harriman in the late 1940s, at a time when rocket designers and others were turning their eyes to space flight and making plans for the technology afforded by the captured V-2s from the Second World War. Getting there was *so* easy in Heinlein's story; unlimited capital plus boundless faith in American ingenuity (and more than a little chicanery) were the engines that powered Harriman's fictional travels.

Heinlein didn't foresee that the Russians would beat the United States to the first human space launch, or that NASA's lunar explorations would end in the early seventies, financially starved by changing politics, new programs, and the recent debacle of the Vietnam War. Certainly by 1978 (the year he set his story) men had walked the Moon, but no one had returned for further exploration or colonisation. There were stations in low Earth orbit, Skylab and Salyut in the 1970s and Mir in the 1980s, and both the US and Soviet Union were sending robotic probes to the planets. However, Harriman's commercial exploitation of Moon trips weren't happening.

At least one of Heinlein's assumptions did come true: commercial interests getting involved in space exploration. They were starting to do so by the late 1950s, although not necessarily in the ways he predicted. In many of his stories, he had corporations paying to do the hard work of settling Venus, doing asteroid mining, building lunar colonies and colonising Mars. If a government was involved, it was as a regulatory partner or through the military.

In reality, space exploration was largely initiated and completely controlled by government agencies such as NASA and the Soviet design bureaus. Aerospace and other companies were paid to develop and supply hardware systems and expertise: electronics, avionics, data, life support, propulsion, space suits, sensors, food and much more. NASA supplied the management, the funding and the astronauts.

Not all the companies involved have been big aerospace concerns focused solely on NASA projects. ILC Industries, which has traditionally made all of NASA's space suits, for example, is a privately held company that specialises in fabrics and materials used in space suits, airships and blimps and protective suits for firefighters. It also provided the airbag assemblies that allowed the Mars Pathfinder and Mars Exploration Rovers to land safely on the Red Planet. Some contractors to NASA have been very small companies, providing services ranging from software development to media production for public educational outreach.

The government/private sector model has held up well and NASA still relies on contractors (and subcontractors) for much of its work. It will likely continue to do so for the foreseeable future. However, the rise of 'commercialisation' in public–private partnerships has given birth to a phenomenon called 'NewSpace'. It is made up largely of newer, younger companies and executives who are part of a growing network of entrepreneurial players practising innovative, disruptive ways of doing space exploration.

Around the world, other private companies are involved in space exploration efforts, with Arianespace and ESA as one example. Although not as old as some US aerospace companies, Arianespace touts itself as the first commercial space transportation company. It formed in 1980 to develop the Ariane family of rocket launchers, and works with the Russians on the Soyuz-2 rocket and the Italians on the Vega rocket. The company has focused mainly on launching communication and weather satellites – more than 550 satellites as of 2017. Arianespace has partners in Germany, Belgium, Denmark, Spain, France, Italy, the Netherlands, Norway, Sweden and Switzerland.

The Chinese government is looking toward private tech companies to help with its aggressive space exploration plan.

The country announced it will triple its spending for both state-run and privately held space corporations. This is a change from the mainly government-run space efforts it has made in the past, although observers say some private actors in the space sector are largely government-owned. One example is Landspace, a 2015 start-up that is focused on building a family of rockets using private investment money to do so.

In the old Soviet Union, design bureaus took responsibility for drawing up plans for spacecraft and missions. Once the prototypes were proven, they were built within the space programme. That has changed, particularly since the collapse of the Soviet Union. Gazprom is a Russian conglomerate concerned mainly with oil and gas production, but it also oversees Gazprom Space Systems (GSS), supplying and operating satellites for communications and reconnaissance. In addition to supporting Russian space activities, GSS also supplies access to broadcast companies and government communications systems. More than 100 private companies work in the Russian space industry, many in manufacturing launch vehicles and rockets as well as satellite development.

Russia also has cooperative agreements with former Soviet republics for space exploration services. ISC Kosmotras is a joint project between Russia, Ukraine and Kazakhstan. It combines agencies and bureaus providing facilities and services. For example, the Russian state space corporation, Roscosmos, handles facility access for launch and training at the Baikonur Cosmodrome, while groups in Ukraine are responsible for the design and manufacture of launch vehicles. Roscosmos itself is the successor to the Russian Federal Space Agency, and the result of the government nationalising much of the space production infrastructure.

The future of space exploration will depend increasingly on such public–private partnerships, or even outright entrepreneurship, such as Elon Musk's private efforts to build a rocket family. His company, SpaceX, now sells launch services to both public and private interests. Private companies could well lead the way to space where governments once did. However, there's another aspect to be considered: public–private partnerships in space have inspired similar collaborative efforts in other fields of science. UNESCO, the United Nations Educational, Scientific and Cultural Organisation, released a report in February 2017 noting a trend toward such partnerships in medical research, climate change mitigation and as a way to foster economic recovery and overcome the limits of austerity budgets in many countries. The role commercial entities play is growing, and often in ways that benefit our society.

Commercial interests in space exploration have had a very real effect on our ability to communicate and learn more about the world around us. As one of the most important results of space exploration, the communications satellite, is a good way to begin our examination of commercial interests in space.

Satellites Lead the Way

In the early days of the Space Age, as both the US and Soviet Union were scrambling (and sometimes lapping each other) in their eagerness to get to the Moon, the ability to communicate over vast distances was critical. As we have learned, the idea for orbiting communications satellites came from Arthur C. Clarke (see chapter 2) and his 1940s publication outlining the use of space-based telecommunications. It took until the 1960s for space-based telecommunications systems to be deployed. Today, there are well over 1,000 such satellites in orbit, deployed by nearly fifty countries through nearly 100 companies and consortia.

The first artificial satellite to be considered a communications relay was Echo 1 in 1960. It was really a kind of balloon, nicknamed a 'satelloon'. It reflected signals from its aluminised surface, like a really big mirror, but for its time it was a big step forward in telecommunications. It was followed by more advanced satellites, such as Telstar, an active, direct-relay satellite owned by AT&T.

The Communications Satellite Act of 1962 resulted in the formation of COMSAT, a public entity with US federal support and ownership by various communications stakeholders. It was involved with the creation of INTELSAT a few years later, an international consortium responsible for improving global communications satellite coverage. Today, COMSAT supplies satellite communications for industry, government and the military.

COMSAT is not the only company in the communications satellite business. UK-based Inmarsat offers telephone and data services to its customers, while the main communications companies in most countries now maintain their own satellites or lease time on commercially available satellites. In particular, Russia uses its own satellite communications company, plus the private Gazprom Space Systems.

In the US, DirecTV (now part of AT&T) offers entertainment programming via satellite to millions of homes and businesses. Sirius XM began as a satellite radio service in the US and now also serves Canada. DISH Network is another direct-broadcast entertainment provider that supplies television, audio programming, interactive TV and internet services. A host of other companies supply similar satellite news and entertainment services to wide areas of the planet.

Secure communications are highly prized by countries able to launch and maintain their own satellites. Whether they are in geostationary orbit over their territories or exist in constellations that see wide areas of the globe (such as the Iridium satellite telephone network), they provide near-instantaneous communications capability to people around the world. This is why so many space agencies and their countries invest in whatever satellite access they can get.

We rely on other satellite systems too, including weather satellites (used mostly by government, but also for commercial weather forecasting), and navigational needs. The best-known of these are the Global Positioning Satellite (GPS) constellation, the Global Navigation Satellite System (GLONASS) in Russia, the BeiDou system in China, and Galileo in Europe.

The first generation of GPS satellites were built by Boeing, with the first ones launched in 1978. The system was also called NavStar and originally was built for the US military as the Defense Navigation Satellite System (DNSS). It remained largely military in focus until the downing of Korean Air Lines Flight 007 by the Soviet Union in 1983. The airliner had wandered off-course due to navigational errors. As a result of the incident, the Reagan administration decided to allow worldwide access to DNSS, although initially not with the precision available to the military. The same signals were eventually made available to everyone, and GPS receivers began to proliferate through the private sector. The GPS system itself is owned by the US government, and maintained by the Department of Defense. It provides timing information used for accurate positioning in navigation, and its signals are used every day by banking networks, cellular phone networks, transport systems and in handheld devices and private cars.

The way GPS works is by signals from the constellation of satellites that orbit 20,000 kilometres above Earth's surface. There are thirty-one satellites, each orbiting twice a day in a manner such that at least three are in view from any point on Earth. Each one transmits its current orbital position so that any GPS receiver can determine its distance to that particular satellite. Once the GPS receiver measures three individual satellite distances, it has enough information to determine its own location on Earth. To maintain the highest accuracy the satellites carry atomic clocks, which provide a very stable signal. Because GPS systems can help determine exact positions, they are useful in other applications. For example, they assist geologists and others in determining just how much a crustal plate has shifted due to earthquakes and other tectonic motions.

Space Commercialisation in the US

The private sector has always been a part of US space exploration, even in the early days. Just as the military and other aspects of government have been dependent on contractors and suppliers, space exploration entities today rely on commercial expertise to supply them.

The various companies involved in the space programme traces a complex system of 'begats'. Many aerospace companies, created from scratch, have spun off others, merged with each other and created new spinoff companies that have been bought out, merged and united with others. For example, the Apollo capsules were built by North American Aviation, which began life in 1928 as an aircraft manufacturer. During the Second World War, the company built combat aircraft. After the war, it branched out to nuclear research, navigation, radar and data systems, and ultimately rocket engines for the space programme. That rocket division was spun off to become Rocketdyne, which eventually became part of Rockwell International (which helped build the space shuttle orbiters). It was then sold to United Technologies and renamed Pratt & Whitney Rocketdyne. That was later sold to Gencorp, and a later merger with Aerojet resulted in a new entity called Aerojet Rocketdyne. Buyouts continue to the present day, with announcements in 2017 of aerospace giant Northrup Grumman's interest in acquiring Orbital ATK. Such mergers are to be expected as the space industry grows and changes.

NASA itself is intricately interwoven with contractor-agency relationships. Many companies work on-site at NASA installations; others work on agency projects in their own facilities, either as prime contractors or subcontractors. For example, United Space Alliance (USA) was contracted with NASA to provide services for space shuttle launch operations. The company is actually a joint venture between Lockheed Martin and Boeing. Until the space shuttle programme ended, USA focused largely on operating the shuttle fleet – taking care of orbiter processing after each flight, readying shuttles for future flights. With the retirement of the shuttle fleet, USA no longer pursues active contracts.

This is not the end of the partnership between Lockheed Martin and Boeing. In 2006, the expendable launch vehicle divisions of each company joined forces to form United Launch Alliance (ULA), which provides launch services using Atlas and Delta rockets to NASA, the military and non-governmental groups. Until 2015, ULA was the sole provider of launch services to the US government. That changed in 2016 when the US Air Force chose SpaceX to launch its satellites and its super-secret X-37b orbiter.

Launch services aren't the only avenues for corporations to participate in space exploration. Satellite design and payload manufacture keeps companies busy. Several Mars mission probes – Odyssey, Phoenix, Global Surveyor, Mars Reconnaissance Orbiter, MAVEN – plus Juno at Jupiter and the OSIRIS-ReX to asteroid Bennu, for example, benefit from work done by scientists and technical experts at Lockheed Martin. During the training for the first Hubble Space Telescope servicing mission, the company (known then as Martin Marietta), built a mock-up of the shuttle for astronauts to use.

Ball Aerospace in Boulder, Colorado, has worked on several of the same missions as Lockheed Martin. It manufactured all of the Hubble Space Telescope scientific instruments, and was key in developing the corrective optics used to 'fix' the telescope's spherical aberration problems. Ball's satellite expertise goes back to the early days of space exploration, with the Orbiting Solar Observatory programme, and was then known as Ball Brothers Research Corporation. Hughes Space and Communications company built the last OSO satellite, plus Syncom, the first geosynchronous communications satellite, as well as ATS-1, the first geosynchronous weather satellite.

It's worth taking a further look at these traditional commercial participants in the aerospace and space exploration game before moving our attention to the era of 'NewSpace'. Launch vehicles (rockets) remain the highest priority for space access around the globe. In the US, the main companies involved in the Space Race began as aircraft and motor manufacturers. Chrysler Corporation became the prime contractor to the US Army Ballistic Missile Agency to build rockets using Rocketdyne motor technology. Led by Wernher von Braun, the combined groups built the Redstone rocket series in the 1950s. Other companies, such as Grumman and McDonnell (now McDonnell-Douglas), were called upon to design and build crew capsules, space station parts and shuttle hardware.

Aerojet Rocketdyne (owned by GenCorp)

Rocket motors and engines are integral parts of space exploration hardware. Aerojet, formed in 1936 by a group of scientists at the Guggenheim Aeronautical Laboratory at CalTech, led by Theodore von Kármán, continues to supply hardware to NASA. Recent contracts focus on propulsion systems and rocket motors for the Orion Multi-purpose Crew Vehicle. It was acquired by GenCorp, which also purchased Pratt & Whitney Rocketdyne (a division of aircraft company Pratt & Whitney, which is itself a subsidiary of United Technologies Corporation) in 2013.

Rocketdyne was originally North American Aviation (NAA), a unit of engineers testing the V-2 rockets brought to the US from Germany after the Second World War. The company quickly moved into the production of its own rockets – the Thor, Delta, and Atlas – and supplied the main engines for the Saturn launchers and space shuttles. The company merged with North American Rockwell in 1973 to form Rockwell International.

During the Space Race, NAA built the Apollo command and service modules. Rockwell International went on to build the space shuttles. Eventually Boeing purchased most of the company's aerospace and defence assets.

Boeing

The Boeing Company has long been involved with the US space programme, and now provides launch services with the Delta II, III and IV rockets through its United Launch Alliance partnership with Lockheed Martin. Currently, the company is developing the Crew Space Transportation capsule for NASA – the CST-100 Starliner, partnering with Bigelow Aerospace. This vehicle can carry up to seven passengers to space and, unlike earlier Apollo, Gemini and Mercury capsule designs, is a reusable craft.

Lockheed

The Lockheed Corporation was responsible for the Atlas-Agena rockets that took robot probes to the Moon, Venus and Mars. The company also provided training for the Gemini training project. Originally an aircraft company started by Allan and Malcom Loughead in 1912, its first project was the model F-1 flying boat. In 1926, the company became Lockheed Corporation and continued producing aircraft for years, despite the Depression, bankruptcy and subsequent buyouts. After the Second World War (for which it produced nearly 20,000 aircraft), the company continued its work on military aircraft, and passenger planes for peacetime uses. During the Cold War it supplied guided ballistic missiles for the US Navy. Lockheed and Martin Marietta merged in 1995 to create Lockheed Martin, and its space systems division remains heavily involved in creating spacecraft and instruments.

Martin

The Glenn L. Martin Company was the contractor for the Titan missile series starting in 1955. It was originally formed as an aircraft company in 1912, supplying both military and civilian needs. Over time, the company moved to guided missile and spacecraft hardware. In 1961, Martin merged with American Marietta to become Martin Marietta Corporation. The merger above, with

Lockheed, produced today's Lockheed Martin company. The resulting company is now building the Orion Multi-Purpose Crew Vehicle for NASA and working on advanced mission concepts for eventual long-duration trips to Mars.

The Commercial Uses of Space

While the first payloads lofted by commercial providers were satellites, there are currently several main avenues for commercial space ventures: cargo and resupply missions to the ISS, crewed missions and space tourism. Several companies working on these come from the Heinlein-style mould of the entrepreneur: people with a singular vision for space-related work who set out to make those visions real.

In the past decade and a half, the shift from purely governmental space access requiring paid contractors and suppliers to private ventures and space entrepreneurship has radically changed the face of space exploration. In the US, the national space policy shifted toward more reliance on commercial space services and products. Under the administration of President Obama, the country heavily pursued the use of commercial launch services.

One of the interesting philosophical developments in the private space sector is the idea of 'new space', or, as it has evolved to be known, NewSpace. Its adherents, influenced by or involved in what has often been referred to as the 'Silicon Valley' style of entrepreneurship, see themselves and their organisations as disruptive to the 'old' ways of doing business in space exploration. Among those considered NewSpace are Richard Branson's Virgin Galactic, Elon Musk's SpaceX and Jeff Bezos's Blue Origins. Interestingly, this strand of ground-breaking, imaginative technology and entrepreneurship also seems to embrace some approaches at NASA; for example, the New Horizons mission and its leadership have been held up as part of the NewSpace movement.

As both new and 'old' space companies innovate in the space sector, they continue to face challenges in creating newer, better, more economical spacecraft. The costs of launch are high, and were formerly borne mostly by government and military budgets. In an era of commercial access to space, companies and governments pursue low-cost ways to loft payloads to space. Certainly creating smaller payloads (for example, tiny nanosatellites weighing less than 10 kilograms and CUBEsat modules, which weigh 1.3 kilograms or less) is one way to reduce launch costs, which can run to thousands of dollars per kilogram. Keeping the weight of the launch vehicle down is another way.

One of the biggest expenses of any launch is the launch vehicle itself. Single-use multi-stage rockets have been the norm. The space shuttles themselves were reusable, as were their solid rocket boosters, but not their main tanks. There is not yet a completely reusable launch system, although there have been experiments with different types of reusable stages and rockets. The early designs of the Delta Clipper, which was an unmanned orbital prototype with single-stage vertical take-off and landing, showed promise during its test flights. NASA budget constraints grounded the programme and engineers from that programme went on to work on the Blue Origin project.

A number of factors converged early in the twenty-first century, including fuel costs and the invention of newer and more lightweight materials for rocket bodies that pointed the way toward reusable multi-stage launchers. SpaceX's Falcon 9 has launched multiple times with a reusable first stage. It lifts off normally; then, after the second stage separates, the first stage manoeuvres its way back down to a vertical landing on land or a barge. The first successful landing was on 21 December 2015. The Blue Origin New Shepard, a vertical take-off and landing rocket, made its first launch past the von Kármán line (the theoretical 'entrance' to space) on 23 November 2015. Reusables are a game-changer, bringing launch expenses down and widening access to orbit.

Future developments for lighter technology include metal-free composites, such as those used by Scaled Composites on its SpaceShip Two. Advanced materials for launch vehicles are one hallmark of the NewSpace attitude, and they can certainly provide benefits beyond the realm of aerospace.

NewSpace thinking is not limited to the US. Its philosophies in the advancement of space exploration (including international trade considerations, worldwide markets and political interests) are shaping the space industries of China, Japan, Europe and other countries. The companies involved (whether they are public–private partnerships or venture capitalist endeavours) are making inroads into traditional space domains. Some are also pushing into markets not traditionally associated with exploration, such as the creation of space 'real estate' (where Bigelow Aerospace expects to be a major player), mining, funeral services, education, arts and culture, and tourism. It's worth taking a closer look at some of these companies and their products and services.

Bigelow Aerospace

At the current time, the ISS and the Chinese Tiangong 2 space stations are the only orbital platforms for research. As more private enterprises and space agencies look to orbital space and beyond for their missions and space tourism, space habitats will become a necessity. Lofting bits and pieces of space stations is a pricey proposition. That's where companies such as Bigelow Aerospace are hoping to make their mark.

Founded by entrepreneur Robert Bigelow (who once owned a chain of motels called Budget Suites), the company is best known for expandable space station modules. These units are made of a tough, multi-layer material called Vectran that expands to create additional living space. The Vectran is resistant to debris strikes and provides radiation shielding. The company's Bigelow Expandable Activity Module (BEAM), a 16-square-metre habitat expansion module, was deployed on the ISS in July 2016. During its first year it was tested for micrometeorite hits, and astronauts continue to study how well it withstands radiation.

In the future Bigelow habitats could be used for transiting scientists on a space station orbiting the Moon for surface operations staging. Bigelow-based commercial space stations (still in the planning stages) could also provide room for other commercial activities, and possibly even accommodate space tourists.

Celestis

While space exploration can be a life-or-death experience, the Celestis company is interested in bringing people to space after they've died. To do that, the company launches small portions of a person's cremated remains to space on rockets going to space for other reasons. The company has sent remains of such notables as

Star Trek's Gene Roddenberry, sixties activist and writer Timothy Leary, astronaut Gordon Cooper and *Star Trek* actor James Doohan, and helped in getting a small portion of the ashes of planetary scientist Eugene Shoemaker to the Moon aboard Lunar Prospector. Celestis dates back to the 1980s, when two companies (Space Services Inc. of America and Celestis Group) – came up with an idea inspired by science fiction stories of burials in space.

RocketLab

RocketLab occupies a special niche, offering cost-effective commercial rocket launches. It began in 2006, founded in New Zealand by CEO Peter Beck. It has since won a government contract from the US to study launching nanosatellites to orbit, and has opened US offices. The company's motto is 'Space is Now Open for Business' and, with its test launch of the Electron rocket in 2017, is moving forward with plans for small satellite delivery to low Earth orbit. This private venture company received major investments from venture capitalists and companies such as Lockheed Martin.

Space Adventures

While the era of space tourism is still very much in its infancy, Space Adventures (founded in 1998), is working on the future of private space flight and tourism. Its current offerings include a space station visit, a chance to become the first private citizen to perform a spacewalk (from aboard a Soyuz capsule), suborbital flight experiences (when such launches become available), space flight training at Star City in Russia, a tour expedition of a Soyuz launch, zero gravity experiences, and – when Moon trips are feasible – a circumlunar expedition aboard a Russian launch vehicle. While most of these are well in the future, they do demonstrate the allure going to space has for some people (even if they will only be available to the well-heeled among us).

Space Ops (Australia)

With nanosatellite launches a reality, companies that can provide low-cost boosts to space have a bright future in exploration. Space Ops is a group of engineers and rocket experts whose rocket,

Above left: Dr Robert H. Goddard and his liquid oxygen-gasoline rocket on 8 March 1926. (Courtesy NASA/Esther C. Goddard)

Above right: A replica of the V-2 at the Peenemunde Museum in Germany. (Courtesy A. Elfwine under Creative Commons 3.0)

Below: The first photo taken of Earth from 105 kilometres altitude by a camera aboard a captured V-2 rocket. The US launched it from White Sands Missile Range, New Mexico. (Courtesy White Sands Missile Range/Applied Physics Laboratory)

Left: The creators of Explorer 1 holding a model of it at the launch press conference on launch day (or the day before), 1 February 1958 (or 31 January 1958). Starting from left side: William Pickering, James Van Allen, Wernher von Braun. (Courtesy NASA)

Below: Dr Wernher von Braun is shown in this photograph from 1961 with members of his management team. (Courtesy NASA)

Above: Ed White performing the first space walk by an American on 3 June 1965. Soviet astronaut Alexei Leonov had performed the first on 18 March 1965. (Courtesy NASA)

Below: The Earth as seen from the Moon during the Apollo 8 mission, the first human mission to circle the Moon. Taken Christmas Eve, 1968. (Courtesy NASA)

Edwin Eugene 'Buzz' Aldrin, walking on the lunar surface during the Apollo 11 mission, 21 July 1969. Neil Armstrong, who took the photograph, can be seen reflected in Aldrin's helmet visor. (Courtesy NASA)

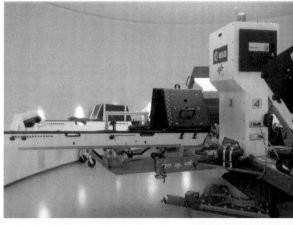

Above: As astronaut candidates, Nicole Mann and Jessica Meir train for extravehicular activity in the Neutral Buoyancy Lab at NASA's Johnson Space Center in Houston. (Courtesy NASA/B. Stafford)

Right: Human centrifuge at DLR in Cologne, Germany used for human physiological tests. The high accelerations experienced during suborbital flights may necessitate testing or even training in human centrifuges to determine if participants are fit for space flight.

Supermoon rising behind the Soyuz rocket poised to take NASA astronaut Peggy Whitson, Russian cosmonaut Oleg Novitskiy of Roscosmos and ESA astronaut Thomas Pesquet to the International Space Station (ISS) on 18 November 2016. (Courtesy NASA/Bill Ingalls)

Above: Space shuttle *Endeavour* on the night before her last launch on 1 June 2011. (Author's collection)

Below: The ISS, orbiting Earth and carrying crew members from various countries on months-long expeditions. (Courtesy NASA)

NASA astronaut Jack Fischer, working aboard the ISS on 3 June 2017, sent back this image, joking that it was like having a corner office in space. (Courtesy NASA)

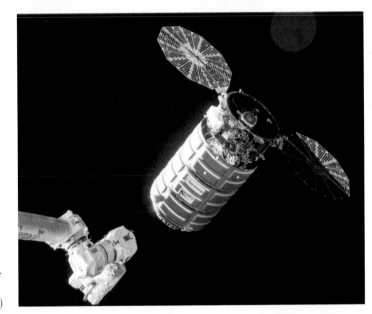

The Cygnus resupply ship from Orbital ATK approaches the ISS on 22 April 2017, before its capture and installation to the Unity module with the Canadarm2 robotic arm. (Courtesy NASA)

The Bigelow Expandable Activity Module, shown attached to the ISS, 17 April 2017. (Courtesy NASA)

Orbital ATK's Cygnus cargo craft (left) is seen from the Cupola module windows aboard the ISS on 23 October 2016. The main robotic workstation for controlling the Canadarm2 robotic arm is located inside the Cupola and was used to capture Cygnus upon its arrival. Expedition 49 unloaded approximately 5,000 lbs of science investigations, food and supplies from the newly arrived spacecraft. (Courtesy NASA)

ESA astronaut Thomas Pesquet is photographed during a spacewalk in January 2017. (Courtesy NASA)

NASA astronaut Tracy Caldwell Dyson looks through a window in the Cupola of the ISS. (Courtesy NASA)

STM Deploy – Students from St. Thomas More Cathedral School in Virginia watch as their STMSat-1 CubeSat is deployed from the ISS in December 2015. (Courtesy NASA)

College students learning the rudiments of space missions during ESA Academy training week. (Courtesy ESA)

Above: Visual Impairment Intracranial Pressure (VIIP) Syndrome was identified in 2005, and a spaceflight risk. Here, NASA astronaut Karen Nyberg of NASA uses a fundoscope to image her eye while in orbit. (Courtesy NASA)

Below: Astronaut and doctor Mae Jemison aboard the Spacelab Japan (SLJ) science module on the Earth-orbiting *Endeavour* space shuttle. Making her only flight in space, Jemison was joined by five other NASA astronauts and a Japanese payload specialist for eight days of research in support of the SLJ mission, a joint effort between Japan and United States. (Courtesy NASA)

The fourth test launch of the New Shepard vehicle on 19 June 2016. (Courtesy Blue Origin)

The landing of the New Shepard first stage to its launch area on 19 June 2016. (Courtesy Blue Origin)

Above left: A launch of the Falcon 9 rocket, built by SpaceX, carrying a capsule of supplies to the ISS, 3 June 2017. (Courtesy SpaceX)

Above right: SpaceX Falcon 9 first stage settles to a landing after deploying a supply capsule to the ISS. (Courtesy SpaceX)

Left: The interior of the Dragon V2 capsule designed ferry up to seven people to Earth orbit. Built by SpaceX in partnership with NASA. (Courtesy NASA/Dmitri Gerondidakis)

Below: The electromagnetic spectrum is the full extent of light that is radiated or emitted by objects in the universe. The spectrum is shown here with a sampling of the observatories and spacecraft that can detect each part of the range. Also shown, everyday examples of objects that also emit or radiate. (Courtesy Chandra X-Ray Observatory)

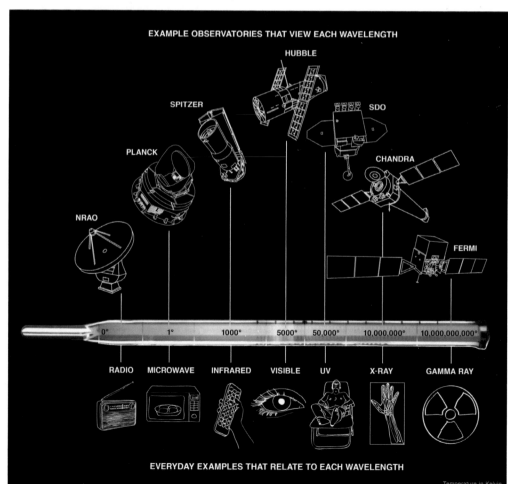

EXAMPLE OBSERVATORIES THAT VIEW EACH WAVELENGTH

HUBBLE

SPITZER

SDO

PLANCK

CHANDRA

NRAO

FERMI

| 0° | 1° | 1000° | 5000° | 50,000° | 10,000,000° | 10,000,000,000° |

RADIO MICROWAVE INFRARED VISIBLE UV X-RAY GAMMA RAY

EVERYDAY EXAMPLES THAT RELATE TO EACH WAVELENGTH

Temperature in Kelvin

This self-portrait of NASA's Curiosity Mars rover shows the vehicle at the Quela drilling location in the Murray Buttes area on lower Mount Sharp. This is a composite of images taken by the rover and combined to give the full view. They were taken around 17 September 2016. For scale, the rover's wheels are 50 cm in diameter and about 40 cm wide. (Courtesy NASA/JPL-CALTECH/MSSS)

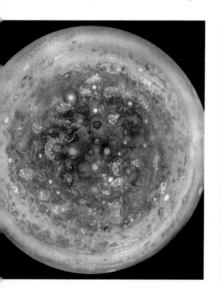
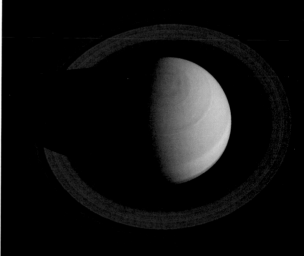

Left: This image shows Jupiter's south pole, as seen by NASA's Juno spacecraft from an altitude of 52,000 km. The oval features are cyclones, up to 1,000 km in diameter. (Courtesy NASA/JPL-Caltech/SwRI/MSSS/Betsy Asher Hall/Gervasio Robles)

Right: NASA's Cassini spacecraft took this image of Saturn and its main rings. The view is in natural colour, as human eyes would have seen it. Saturn sports differently coloured bands of weather in this image. (Courtesy NASA/JPL-Caltech/SSI/Cornell)

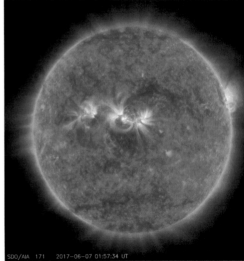

Above left: NASA's New Horizons spacecraft captured this high-resolution enhanced-colour view of Pluto on 14 July 2015. Pluto's surface sports a remarkable range of subtle colours, enhanced in this view to a rainbow of pale blues, yellows, oranges and deep reds. Many landforms have their own distinct colours, telling a complex geological and climatological story that scientists have only just begun to decode. The image resolves details as small as 1.3 km across. (Courtesy NASA/JHUAPL/SwRI)

Above right: A Solar Dynamics Observatory view of the Sun on 6 June 2017, showing plasma moving along magnetic field lines. This is an extreme ultraviolet view of the Sun. (Courtesy NASA)

Below: This artist's concept shows what a system of exoplanets (planets orbiting another star) called TRAPPIST-1 planetary system may look like, based on available data about the planets' diameters, masses and distances from the host star. The system has been revealed through observations from NASA's Spitzer Space Telescope and the ground-based TRAPPIST (TRAnsiting Planets and PlanetesImals Small Telescope) telescope, as well as other ground-based observatories. The seven planets of TRAPPIST-1 are all Earth-sized and terrestrial, according to research published in 2017 in the journal *Nature*. TRAPPIST-1 is an ultra-cool dwarf star in the constellation Aquarius, and its planets orbit very close to it. (Courtesy NASA/JPL-Caltech)

Above: The Carina Nebula is an extensive cloud of gas and dust visible from Earth's Southern Hemisphere. It lies at least 6,500 light-years from Earth. It has areas of star birth and at least one massive star called Eta Carinae that may explode as a massive supernova (called a hypernova). This image of the nebula was taken with the Hubble Space Telescope (HST). (Courtesy NASA/ESA/STScI)

Below: In 2004, the HST observed a tiny part of the sky for ten days, capturing views of very distant galaxies. This image shows the same view taken in 2009 with a newly enhanced camera. Other 'deep field' images taken since then, aimed at different parts of the sky, show there are galaxies in every direction of the sky, as far as can be detected. Some of the galaxies are among the most distant ever observed. (Courtesy NASA/ESA/STScI)

Mars missions will someday take humans to the Red Planet. This artist's concept shows what a future science team might be doing there, and some of its vehicles and work modules. (Courtesy NASA)

The Martian environment is very harsh and human crews will have to protect themselves against the temperature extremes, thin atmosphere and high radiation levels. This vehicle is a concept being tested that combines a rover with suits that the Mars-nauts could climb into once they arrive at a study area. (Courtesy NASA)

Rocky 1, will begin flying by the 2020s. The company's major investments come from across Australia.

Space Mining

Of all the futuristic directions that space exploration could take, space mining is one of the most discussed. The idea is to mine asteroids for their water deposits, as well as other raw materials – gold, iron, cobalt, iridium, silver, nickel, molybdenum, oxygen, water, titanium, and tungsten. Some of these materials would be used in space as construction materials for craft, habitats, and for rocket propellants. Others would be mined for use on Earth, particularly the precious metals.

Why mine asteroids? It's well known that some of the minerals carried in asteroids are in short supply on Earth (or will be). Some experts suspect we could exhaust Earth's supplies of some of the rarer ones before the end of the twenty-first century. Asteroids could resupply these valuable elements. Planetary scientists have long known asteroids have the same basic mineralogic profile as early Earth (and, indeed, seeded the newborn planet). 'C-type' asteroids and 'blown-out' comets are likely abundant in minerals and water, making them especially valuable.

Such *'in situ* resource utilisation' offers not just commercial rewards but advances in the scientific study of asteroids. Planetary scientists could study them to gain more insight into the materials that existed in the early solar system. These primordial grains eventually accreted to form asteroids, planetesimals and, eventually, the planets. Those studies would almost certainly include a search for any evidence of the chemical precursors of life.

Asteroid mining will require carefully selecting the right targets and then devising methods to get to them. It will not be an overnight process. Those who want to proceed with it will have to decide whether to process the materials on-site; alternatives include moving the asteroids closer to Earth for mining (an idea proposed during President Obama's term of office but recently rejected by the subsequent administration), or even depositing the asteroid in orbit around the Moon, where it can be exploited. The technology for such operations has yet to be invented, although many plans are certainly in embryo. This will also require launch capability as well as crew habitat and transport.

Deep Space Industries

In the first decades of the twenty-first century, several companies and at least one country have staked their claim on the space mining business. Deep Space Industries proposes using the Prospector-1 spacecraft to search out likely mining targets among near-Earth asteroids. The company proposes to use water as a propellant for the spacecraft. Camera systems would help locate and study the target asteroids, and the spacecraft would communicate its findings back to Earth.

The Prospector-1 craft would be the first commercial interplanetary mission. The plan is to rendezvous with a near-Earth asteroid. After a brief survey period, it would land and analyse the surface materials using on-board instruments. Harvesting missions would next bring materials back to near-Earth space for processing into fuels, drinking water or even building supplies for spacecraft or habitats. A NASA Small Business Innovation Research grant in 2015 united the company with Florida Central University to study the development of propellants made from asteroid 'regolith' (a surface covering of dust, sand and pebbles). The company is partnering with the government of Luxembourg, which announced in 2016 its intention of laying the legal and technical groundwork for mining and the ownership (and utilisation) of asteroid resources. The country drafted a law to ensure companies would own the resources they mine, and has pledged major investments.

Planetary Resources: The Asteroid Mining Company

The private venture-funded company Planetary Resources is another to set its sights on asteroid mining. It was founded as Arkyd Astronautics in 2009. Like Deep Space Industries, Planetary Resources has also partnered with the country of Luxembourg, and announced plans for a series of robotic mining missions. Luxembourg is not a space power, but is a centre of industry in Europe. It has largely made money from the steel industry, and has long been involved in the communications industries. In November 2016, it passed a law allowing private companies to keep resources they find in space. Planetary Resources has been developing mining technologies that could be used on asteroids. No date is set for their first venture, although the company secured funding late in 2016 to proceed with development of a constellation of satellites to observe Earth.

Mining the Moon

Utilising lunar resources has often been cited as a way to get low-cost building materials outside of Earth's gravity well. The Chinese space agency's intention to build a lunar village may well include plans for mining the resources to sustain the colony, including water ice (valuable for life support and propellants). At least one NASA mission, called Resource Prospector, is being discussed and could be the first mining expedition to the Moon. It would study the lunar poles, where water ice deposits have been found in the past by the Lunar Crater Observation and Sensing Satellite and Lunar Reconnaissance Orbiter missions. If future lunar colonies and research stations are to be the norm, they will need to make their own air and water, rather than ship them from Earth at very high expense. Hydrogen and oxygen (the basic components needed) are known to exist on the Moon, particularly in the solar polar regions. To mine and process them into consumables would require technology that would be shipped ahead for the inhabitants to use. If missions to Mars are to happen, lunar mining may well provide the raw materials needed for making spacecraft propellants for trans-Mars ships.

NASA's Resource Prospector is typical of similar missions under consideration. It would deploy a rover on the lunar surface to excavate hydrogen and oxygen (called 'volatiles'), and continue the search for water. The mission could launch to the Moon within a half-decade and provide a good test of the concept of *in-situ* resource utilisation.

At least one private company has won US government approval to proceed with a lunar mission of its own. A closely held entrepreneurial venture, Moon Express is vying for Google's Lunar X Prize. This is a contest sponsored by Google to award USD $20 million to the first team to land a robot on the Moon, travel 500 metres on the surface and send images and data back to Earth.

The current US administration has expressed great interest in resource extraction on the Moon. However, there are legal and treaty considerations to be discussed, since strip-mining the Moon for private profit (as opposed to future mission needs) would violate the 1967 Outer Space Treaty. In fact, most mining operations on the Moon (and asteroids) will need legal precedents worked out.

A Mixed Future: Old Space vs NewSpace

Debate about whether 'new space' or 'old space' is the future for commercial space exploration continues. There are those who argue for broader definitions of commercial space activities that also embrace traditional companies long associated with space activities. These include companies serving primarily government customers, such as Arianespace, the Boeing-Lockheed Martin/United Launch Alliance (ULA) partnership and others such as DigitalGlobe, which provides remote sensing services to a wide array of customers. Some argue these are not part of NewSpace movement; yet they are still part of the commercial space sector, even if they do receive most of their income from government contracts.

The mix of the NewSpace and 'old space' commercial interests has created new partnerships when it comes to servicing NASA and other space agencies. While NewSpace likes to think of itself as daring, there are still government contracts for them (and others) to pursue.

At the end of the space shuttle era, NASA was looking for ways to service and resupply the ISS. It was directed to look to the private sector, which opened the gates for missions through its Commercial Orbital Transportation Services (COTS) programme. Although the agency has contracts with Roscosmos to supply crew and access to ISS, it was important to build and maintain a US-based supply chain to bring food, water, clothing, personal items and cargo to the station. Contracts were awarded to Space Exploration Technologies, better known as SpaceX, and Orbital ATK.

SpaceX is a well-known name today, but has been around for less than a decade. It was formed by entrepreneur Elon Musk, who co-founded PayPal and Tesla Motors. The company's main space products are Falcon launch vehicles, Merlin rocket engines and the Dragon delivery capsules. The first commercial spacecraft to dock at the space station was launched by SpaceX in 2012. The Dragon can both supply the station and return used material to Earth. The Falcon 9 rocket series has had several launches of orbital payloads and successfully returned the first stage to vertical soft landings on either a barge or a land-based target.

Orbital ATK is the product of a merger between Orbital Sciences Corporation (a long-time NASA contractor) and units of Alliant Techsystems (a defence and systems contractor). The company builds satellites and the Cygnus spacecraft, another delivery vehicle to and from the ISS. Both Orbital ATK and SpaceX run regular resupply missions.

In 2016, NASA added a third company to the list: Sierra Nevada Corporation. Its shuttle orbiter design, called the Dream Chaser, cannot run crewed missions, but is configured purely for resupply missions. It is very similar in design to previous space shuttles, but instead of launching with two solid-rocket boosters it will go to space atop an Atlas V, Ariane 5 or a Falcon Heavy rocket. Once its mission is complete, the Dream Chaser will de-orbit and land horizontally, just as the shuttles did. It should be possible to reuse it multiple times. The company has also attracted interest for its spacecraft from ESA.

Commercial Crew Programs and Launch Vehicles
Since the end of the space shuttle missions in the US, the only way for crews to get to space is through Roscosmos and its Soyuz systems. This is, as the US has found out, fairly pricey; the Russians are charging upwards of $81 million per seat. NASA has been developing the Space Launch System (SLS) and various crew capsules. These are not expected to fly for a while yet, which necessitates the reliance on Roscosmos for trips to the space station.

NASA created the Commercial Crew Program (CCP) in 2010 to facilitate the design and construction of human-rated spacecraft to restore the US access to space. The agency selected SpaceX and Boeing for the work. SpaceX is working on making the Dragon capsule a crew-rated spacecraft. Boeing is, as mentioned above, building the Crew Space Transportation Starliner spaceships.

SpaceX is not the only game in town (at least in the US) for launch vehicles. Blue Origin is a privately funded aerospace venture created by Amazon founder Jeff Bezos. The company has run test launches of its spacecraft, and will be supplying upper-stage rockets for United Launch Associates' Vulcan orbital launcher. Its New Shepard boasts a crew capsule atop a booster. It is named for astronaut Alan Shepard. The company's New Glenn system is planned to be a multi-stage launch vehicle, with a reusable first stage and an orbiting crew capsule. Like the SpaceX Falcons, the Blue Origin systems are designed for vertical take-off. After separation, the first stage returns to Earth in a 'tail down' landing, just as the SpaceX Falcon does.

With SpaceX and Blue Origin, the age of reusable launch vehicles is in full swing. However, not all trips have to be lofted by rockets, or even get all the way to space. World View is looking toward what it calls the 'stratospheric economy', offering its balloon-based Stratollite and Voyager flight platforms for flights high in Earth's atmosphere. It is already flying payloads, including instruments for communication, weather, remote sensing and science research, and will eventually cover trips to 'near space' by private individuals.

Taking Regular Folks to Space

Space tourism has been a dream for many decades. Sir Arthur C. Clarke, in his science fiction novel *Imperial Earth*, has his protagonist traveling from Saturn to Earth for a business trip, along with fellow passengers making pleasure journeys much as people enjoy ocean cruises today. Tourists make appearances in some of Heinlein's books, including *Starman Jones*, the story of a young man who becomes an 'astrogator' on a huge space liner.

It's no surprise space tourism would beckon to private entrepreneurs. Still, it's a risky business, and as we've learned elsewhere, sending anything to space, whether cargo or humans, is expensive. A few 'pleasure' travellers have been to orbit through Space Adventures of Virginia; they went on Russian launches and paid between $20–40 million each. They were Dennis Tito, Mark Shuttleworth, Gregory Olsen, Anousheh Ansari, Charles Simonyi, Richard Garriot and Guy Laliberté. Those personal space flights have ended for the time being; astronaut seats for trips to the space station have been deemed more important. However, Blue Origin and Virgin Galactic have been taking applications and deposits for future flights. SpaceX has announced it will fly two tourists around the Moon aboard the Dragon 2 capsule, possibly as early as 2018.

The safety aspects of space tourism will require expanded regulatory structures. Just as transportation on Earth (planes, trains, boats, cars) is tightly regulated to assure safety, space flight for private citizens will need to be as well. Right now there are few regulations in place, and even though they are not often welcomed by private companies, most see the need for at least rudimentary protection. It's not clear Earth-based regulatory bodies have the authority to impose such rules, so it may end up on the shoulders of such companies as Blue Origin and others to write the rulebooks regarding tourist safety in space.

Virgin Galactic, which bills itself as the first commercial space line, was founded by Sir Richard Branson in 2004. He went on to form the Space Company in 2005 with aerospace expert and pilot Burt Rutan (founder of Scaled Composites). The idea was to build commercial space ships. Its SpaceShip Two was introduced to the public in 2009, at the Mojave Spaceport in California. At the time, the company had sold 300 bookings aboard its ships at a cost of $200,000 each to take fliers on short trips. Unfortunately, in 2014 the company suffered a setback with the crash of its VSS Enterprise, which killed of one of the two pilots and forced the company to revamp its design and retool its efforts for SpaceShip Two.

Research Institutes and Non-profits

Independent research institutes focused on space science and research are a growing part of the commercial space sector. Many receive funding from governmental agencies as well as private, often tax-deductible investments, but they are largely entrepreneurial in their outlook and practices. Notable examples in the US include Planetary Science Institute and Southwest Research Institute – one of the oldest independent, non-profit applied research groups. PSI is funded by a mixture of public and private partners, and its scientists are involved in numerous missions for NASA and others. It is the largest non-public employer of planetary scientists in the world, spread across the US and eleven other countries.

SwRI has several divisions that serve both government and industrial applications. The group began as a research institute in Texas in 1947 and has grown through the years to perform research and services in the fields of health, biomedical, chemistry, materials sciences, computers, defence, earth and space, energy, environment, manufacturing and transportation. SwRI partnered with the Applied Physics Laboratory at Johns Hopkins University on the New Horizons mission to Pluto.

The Max Planck Society in Germany is another private research organisation supported by industry partners and donors. The society (which has labs in the Europe and the US) performs high-risk research into space sciences and astronomy that government funding does not always cover. The society was involved with the Rosetta mission, and its scientists have made seminal contributions on other missions as well.

Many such institutions are free-standing, not necessarily affiliated with universities or governments. Some maintain active outreach programs for teachers, students and science museums.

Space Investment

Investments in the space 'sector' of the world's economy are growing. In 2017, the space economy was estimated to be worth $329 billion; it could well grow to be a trillion-dollar investment opportunity soon. Space development, whether for continued telecommunications and research, mining or tourism, could be one of the hottest start-up investments to come along. That's just in the near term, with missions to near-Earth space and the Moon. As with any daring venture, however, there are downsides. Space missions are incredibly expensive and do not always run on time or on budget. Exploration itself is a risky business, both to human life and to an investor's portfolio.

The stakes get bigger when people start talking about Mars missions and human colonies in near Earth orbit. Trips to the Red Planet will be highly expensive, technology-driven multi-national efforts. The advancements will require not just technological investments, but very focused work on the legal and social frameworks to assure all partners get the most science and investment bang for their participation.

Although the current focus of attention for some of these missions may still be in the US, as we've seen with Luxembourg's interest in mining and India's interest in sending people to Mars, the space 'market' is global. As mentioned, China is expanding its space interests very rapidly and is a player to watch, both technologically and financially. If its move into renewable energy is any indication, the Chinese appetite for space exploration could mount a significant challenge to the interests of other nations.

The UK is certainly positioning itself in the NewSpace sector, and other countries are making financial investments – not just in space exploration, but also the necessary science, technology, engineering and mathematics education for their citizens. The coming decades in space exploration will certainly be exciting ones, whether the advances come from 'old' space or 'new'.

One aspect of space exploration that rarely gets mentioned but is absolutely essential is insurance. A launch failure almost always raises the question, 'Who's going to pay for this loss?' In these days when commercial space interests are seeking to protect investments as well as replace equipment lost in a mishap, it's not unreasonable to ask. Space insurance, while not new, is a large part of all space business.

The downside can be pretty bad: a SpaceX launch mishap in 2015 destroyed $110 million in supplies headed to the ISS. Premiums to cover launches begin at $50 million currently. Underwriters look at the type of payload and the rocket being used to send it to space before setting a premium for a given launch.

Insurance makes sense – it's there to cover a loss, whether it's for individuals in a car accident or satellite owners who lose their payload when a rocket blows up. The types of coverage for space launches are similar to those found in aviation insurance, which covers airliners and private planes. They cover loss or damage that could occur to the satellite as it is being built and tested, during its transportation to the launch site, and mating to the rocket that will carry it to space. Then, there's the coverage for the launch period, to pay out if something happens to the satellite during lift-off, while yet another policy can be used to protect against loss while the satellite is in orbit. There are also policies to cover damage and loss when a piece of space debris or a satellite falls to Earth. Companies such as the venerable Lloyd's of London offer launch policies.

Moving Ahead

The commercialisation of space may well initiate new space races that could dwarf the accomplishments of the first one back in the 1960s. Driven by venture capital and enthusiasm, backed by expertise and new technology, the business of space is already making the transition from public to private. Already, as we've seen, NASA is passing the torch to new companies that are taking over launch contracts and vehicle developments. That is changing the way companies do business, moving from long-term development contracts to a short-term focus on developing a payload to be launched on a commercial rocket such as the Falcon 9.

From Telstar I (the first commercial satellite flown in space) to asteroid mining in near-Earth space, commercial interests have always been involved in space. The evolution from government-based projects to the coming age of businesses in space is a healthy process, and gives our spacefaring civilisation more access to space and its resources.

7

Planetary Science: Robotic Steps to the Solar System

We shall not cease from exploration, and the end of all our exploring will be to arrive where we started and know the place for the first time.

T. S. Eliot

Earth is the cradle of humanity, but mankind cannot stay in the cradle forever.

Konstantin Tsiolkovsky

When you look at the stars and the galaxy, you feel that you are not just from any particular piece of land, but from the solar system.

Kalpana Chawla

Solar system exploration is one of the great success stories of the Space Age. Today, planetary probes allow global teams of observers to implement complex missions that can fly by, orbit and land on distant planets, comets, moons and asteroids. Back on Earth, scientists analyse and explain the data and images they supply, sharing their discoveries with the world.

The Excitement of Planetary Exploration
On 15 July 2015, a team of scientists, joined by friends, family and millions of TV viewers around the world, waited to receive the first signal back from the New Horizons spacecraft after its

closest approach to Pluto. The anticipation was high. It was the first time any spacecraft had been anywhere near this most distant planet. Everyone waited in tense silence as Mission Operations Manager Alice Bowman polled the spacecraft team for progress reports. She waited patiently, listening on a headset as engineers checked for any sign of information from New Horizons, more than 4.7 billion kilometres from Earth. From that distance, the signal had to travel four hours and twenty-five minutes between planets. For twenty-two hours, New Horizons had been out of contact with Earth while pointing itself toward the planet. In that time there was a small chance it could collide with a space rock or suffer some other mishap. Only the return of telecommunications telemetry from New Horizons would tell if it had survived the flyby.

Moments passed. The speakers broadcast only a hiss. A few electronic beeps and boops could be heard from computers. Everyone in the control room sat quietly, staring at their computer monitors, willing the spacecraft to 'phone home'. Honoured guests in a nearby auditorium, watching the control room via NASA TV, held their breath in nervous anticipation. Many in the room had been part of missions that didn't phone home, and knew the heartbreak. Would the world hear from New Horizons?

Finally, a message crackled on the crew communications channel. Bowman listened intently, let out a big breath, and raised her head to look at the crowd. 'Okay, copy that,' she said. 'We are in lock with telemetry with the spacecraft.'

New Horizons was back in contact!

Spontaneous shouts and applause broke out across several rooms at the Johns Hopkins Applied Physics Lab control centre in Maryland, the facility where the spacecraft teams and their friends were sequestered. Engineers began reporting positive data from all instruments. Bowman turned to principal investigator Dr Alan Stern and made a final announcement: 'PI, MOM on Pluto 1; we have a healthy spacecraft, we've recorded data of the Pluto system, and we're outbound from Pluto.'

It was a moment of hugs, clapping, cheering and pure human joy – a celebration of years of planning, building, launching and transit to a distant world. Mission accomplished.

That headlong flight past Pluto was actually the end of what Stern called the 'initial reconnaissance of the solar system', an accomplishment tracing its roots to the very earliest days of the Space Age.

New Horizons continues on its way out through the Kuiper Belt. That tiny spacecraft and others are a vital part of space exploration. They extend our eyes and ears beyond Earth to help us understand our neighbourhood in space. Today, scientists from countries around the world continue to use orbiting and flyby missions to explore the Moon, Mars, Jupiter, Saturn, asteroids, comets, the Sun and, of course, Earth.

The Age of Planetary Reconnaissance

Visiting the planets and probing their mysteries is an age-old dream come true. Long before the first spacecraft headed to the planets, going to them was the stuff of fantastical tales. In the late 1800s, satirical stories told of travel to other worlds; mostly to make allegorical social commentaries on the human condition here on Earth. As soon as people realised these points of light in the sky were actual worlds, they let their imaginations run wild, festooning them with steamy swamps and exotic lifeforms. However, beginning in the early 1900s, writers started using scientific concepts borrowed from astronomy and the Industrial Revolution to spin stories of exploration. They imagined a time when men and women took brave steps onto new worlds, conquering alien races along the way.

Rarely did early SF writers foresee the reality of today: planetary exploration done by robotic probes. As we've seen, only a few humans have set foot on another world. Human exploration of the other planets may happen, and it seems pretty likely people will return to the Moon, perhaps within the next few years. For now, what we know about the solar system comes to us across immense distances, transmitted by robotic probes to scientists waiting on the ground. They analyse that data to put together explanations of the origin and evolution of the Sun, Moon, planets, asteroids and comets. In this chapter we'll focus on the means by which planetary scientists study the planets *in situ* using spacecraft, and also look at highlights of their explorations. (To see a complete list of planetary exploration missions, see Appendix D.)

Planetary exploration is, at its heart, a very human endeavour. It's about people reaching out, about mankind's propensity to discover, to learn – and celebrate when we discover something new, as the scientists did during the New Horizons flyby.

Scenes like the one described above have played out dozens of times over the past five decades: anxious scientists on Earth waiting to hear back from a distant spacecraft on a remote world. Team members who have worked for years all face this moment of truth, waiting for a signal from their spacecraft saying, in essence, 'Hello – I made it!'

For most teams it's a time of triumph, a vindication of years of research, engineering challenges, budget battles and lost family time while shepherding a spacecraft to another world. For others, it's tragic, as spacecraft were lost, stopped working or never made their goals. More than one scientist who lived through such an tragedy described it as like losing a family member. Though each scientist faces the chance their spacecraft might not succeed, it's immensely rewarding when they are successful.

Why Study the Planets?

Planetary scientists divide the solar system into several regimes. The inner solar system has the rocky worlds Mercury, Venus, Earth and Mars. The realm of the giants contains the gas and ice worlds Jupiter, Saturn, Uranus and Neptune. Beyond them lies the Kuiper Belt, a region of dwarf planets and short-period comets. The outermost region (which could stretch out to almost a light-year away) is called the Oort Cloud. It contains swarms of cometary nuclei. Between Mars and Jupiter lies the Asteroid Belt, a circumstellar disk of small, irregularly shaped rocky bodies. Asteroids also roam in the space between planets, occasionally crossing their orbits.

The main impetus for solar system exploration is curiosity – to learn what those other worlds are like, how they formed and how they are different from or similar to each other. To be sure, the search for life is also on everyone's minds, whether in the form of microbes on Mars or something exotic thriving in the deep oceans of Jupiter's moon Europa or Saturn's moon Enceladus. Whatever information scientists can glean from the other worlds of the solar system can only benefit our understanding of our own place in the cosmos.

Planetary scientists study the solar system with both ground-based and space-based instruments. For example, the planets Uranus and Neptune, both explored by the Voyager 2 spacecraft as it swept past in the late 1980s, continue to be targets for observatories around – and above – the world, including the Gemini and Keck Observatories and the Hubble Space Telescope. While it is unlikely these two worlds will be visited by spacecraft in the foreseeable future, since there are no missions funded to do so, continued observations from Earth give planetary scientists data about how they change over time.

The same is true of our long-term studies of Mars using ground-based and space-based observatories and probes. Exploration of that planet has uncovered solid evidence of water in its past, something that could bode well for the discovery of life there, past or present. It is also giving future explorers a good look at the territory ahead before they even arrive.

The study of solar system objects may have begun with planets, but in recent decades probes to comets and asteroids have unveiled their secrets too. The same information that tells a scientist the chemical makeup of a comet's ices or the water content of an asteroid will inform companies that want to engage in future mining efforts. Spacecraft data have already unveiled a treasury of information about the solar system's early history.

Remote exploration of planets, asteroids and comets is one of the great success stories of our age. Let's look at what nearly seven decades of space-based planetary exploration has taught us about our solar system.

Sending a Probe to Another World

Scoping out another world requires a spacecraft that can do pretty much what a human explorer can do when taking data and observing a planet, but without risking that person's life in space. Cameras are our eyes. They provide detailed images of planetary surfaces and atmospheres, ring systems and moons. Other instruments measure gases in an atmosphere, sense a magnetic field, characterise a world's gravity, image its moons (if any), then communicate the data to Earth via radio antenna (which itself can function as a science experiment). Most probes have star trackers for navigation, plus other stabilising equipment to keep the spacecraft on target. It all requires electronics (wiring, circuit boards, controllers and so on), plus power – supplied by radioisotope thermal generators and/or solar panels.

The probes don't do this work on their own. They are operated by teams of people back on Earth, sometimes numbering in the thousands if you count all the scientists, technical staff, communications personnel and managers. Beyond them are the manufacturing teams, agency sponsors, and others whose work enabled the probe in the first place. It's safe to say that they're all involved in the spirit, if not the science, of the mission.

Getting a Mission off the Ground

Solar system exploration missions are complex projects. The science objectives drive the selection of instruments loaded on the spacecraft. There are often budgetary constraints and political considerations. There are always orbital constraints – that is, a given mission has to launch in a certain 'window' to reach its target in an economical and timely way.

Once the main mission is outlined, a space agency invites the scientific community to submit proposals for the mission. Teams from private industry, the academic community and even the agency itself are formed. They often have an international flavour. For example, the Rosetta mission, 'Europe's Comet Chaser', was commissioned by ESA. It was ultimately built by prime contractor Astrium Germany, with subcontractors in fourteen European countries and the United States. The science team comprised hundreds of scientists and technical experts from around the world.

Competition for missions is often quite intense; it's not unknown to have dozens of teams working on proposals and appropriate budgets – millions or more likely billions in any currency. Eventually (years later), one proposal and its budget is accepted, and the team goes to work.

The spacecraft is designed, built, tested, retested and readied for launch. These days, interplanetary missions leave Earth atop a rocket. For example, the New Horizons spacecraft was mounted on a high-powered Atlas V 551 rocket. It was a three-stage vehicle with five solid rocket boosters. With all that firepower, New Horizons launched at a speed of 58,000 kilometres per hour. That made it the fastest spacecraft ever to be launched and the first to be sent directly into a solar escape trajectory.

Not every planetary mission needs such a powerful launch boost. Generally speaking, most are launched into orbit around the Sun first, and then go through what's called a Hohmann transfer orbit manoeuvre, using small thruster firings to move the spacecraft from one orbit to another until it reaches a point where it can be sent on to its target.

The Mars Curiosity mission, sent to space atop a United Launch Associates two-stage Atlas V-541 vehicle, used the first stage to get off Earth. The second stage then fired twice – once to get the spacecraft into low Earth orbit and then a second time to send Curiosity on towards Mars. The probe made a series of course corrections on its way to the Red Planet. Before arrival, it decelerated at just the right time to get into the proper trajectory to achieve Mars orbit. To get to the surface, the lander (protected by a curved aeroshell that functioned as a heat shield) entered the atmosphere. Mars's thin blanket of air wasn't enough to slow it down, and the spacecraft deployed a large parachute. That brought the speed from 1,600 km/hour to 320 km/hr, and the heat shield was jettisoned. Eventually, the spacecraft was lowered to the ground using a Skycrane – a hovering machine with four steerable engines that carried the rover to its landing point.

Since all solar system bodies are moving in their orbits, there's no such thing as simply aiming a probe to go straight to its target. The idea is to aim at where the planet will be at a given time. Launch is planned with that in mind, as are the inevitable mid-course corrections and manoeuvres.

Sometimes a spacecraft will get a 'gravity assist' by swinging around another planet and using its gravitational influence as a slingshot to pick up speed and change course as needed. The Voyager mission to the outer solar system is a good example. Both spacecraft picked up a gravity assist at Jupiter and then headed to Saturn. Saturn's gravity 'kick' took Voyager 1 on a trajectory out of the plane of the solar system. Voyager 2's gravity assist altered its trajectory so it could go on to Uranus and Neptune.

During the flyby of a planet, instruments are engaged and data is stored for later transmission back to Earth. The idea is to get a lot of information quickly and start sending it back as soon as possible. Depending on the needs of the mission, many parts of the spacecraft might go into a low-activity mode or hibernation until the next target (if there is one). Others stay active and gather data constantly. When the Voyager 2 spacecraft made its closest approach to Saturn on 25 August 1981 it had been imaging the system for months prior, and some of its instruments studied the interplanetary medium and probed for Saturn's magnetic field and gravitational influence. New Horizons spent much of its time hibernating on the way to Pluto, waking up occasionally for scheduled tests and course corrections. After the close flyby, the spacecraft began sending back its data, taking nearly sixteen months to transmit it all.

Once the principal science teams for a planetary mission have finished their analysis, much of it is eventually made available to other researchers, or to the public via websites. All the information contributes to an ever-growing common store of information about the solar system.

When Planetary Probes Fail
We live in an age when sending spacecraft throughout the solar system seems like 'old hat'. As we'll see below, many missions have been quite successful and that leads to a certain amount of assurance that missions never fail. In reality, however, planetary exploration is a tough business. For all the missions that go well, there are those that don't. During the early years of the race to the Moon, both Soviet and American spacecraft were lost due to launch failures or problems on orbit. Some spacecraft flew off and were never heard from again. More than half of all Mars missions have failed.

For example, in 1990, the Mars Observer probe fell silent just as it was about to arrive at the Red Planet. Its loss was a painful experience, not just for NASA but for the scientists who spent part of their careers planning and building the spacecraft. Ultimately, an investigation showed Mars Observer's fuel system leaked, causing catastrophic spinning and loss of power. The Mars 'jinx' continued with other missions, most notably NASA's Mars Polar Lander and the Beagle 2 (from the United Kingdom). The polar lander crashed into the surface of Mars on its final descent due to some faulty sensors which caused the descent engines to shut down 40 metres above the surface. Instead of settling gently down, the spacecraft crashed.

The UK's Beagle 2 actually landed safely on the Martian surface, but not all its solar panels deployed properly. They ended up blocking the communications antenna and preventing the lander from contacting Earth. It was an unfortunate occurrance that brought its principal investigator a great deal of undeserved criticism.

In 1986, mission planners watched as the Spartan-Halley mission lifted off the launch pad at Cape Canaveral aboard the space shuttle Challenger. It was the culmination of years of planning by scientists and students at the University of Colorado. The spacecraft was to take ultraviolet spectra of the tail of Halley's Comet during its perihelion passage. Unfortunately, seventy-three seconds after lift-off Challenger exploded, killing seven astronauts and destroying the mission.

Even otherwise successful spacecraft can develop problems. The Voyager 2 team had to contend with a balky scan platform as the spacecraft swept past Saturn in 1981. Technical staff were able to devise a workaround to diagnose the problem with the gears and created a fix to get to the platform that held the cameras and photopolarimeter (a specialised camera that photographed Saturn and its rings and measured the polarised light coming from them), and allowed them to be rotated into place as needed.

In 1991, the Galileo spacecraft's high-gain antenna failed to properly unfold, leaving the spacecraft looking like a permanently half-opened flower. The problem grossly affected the probe's ability to communicate its data back to Earth. The mission was saved when mission controllers were able to switch all communications to the low-gain antenna. This reduction in bandwidth slowed the data return rate and affected how much information could be sent back to Earth. Scientists were able to meet about 70 per cent of the mission's primary goals, and counted it as a success.

The New Horizons mission to Pluto survived a few glitches of its own, but thanks to constant vigilance and rehearsals the technical staff was able to diagnose the problems and return the spacecraft to normal operations relatively quickly. The most amazing thing about such mid-flight rescues is that the teams are making these fixes while their spacecraft are speeding away from Earth, millions or billions of kilometres out in the solar system. Despite all the technical achievements represented by modern spacecraft, space flight and exploration constantly reminds us that problems can occur.

Target: The Moon

The idea of sending robotic probes to solar system objects is not a new one. Certainly as far back as Hermann Oberth's early years in designing rockets and writing about space travel in the 1920s and 1930s scientists had ideas about using small spacecraft to orbit and even land on the Moon, Mars and Venus. Oberth himself was inspired by the famous Jules Verne story *From the Earth to the Moon*, about a fictitious Baltimore Gun Club whose members built a 'space gun' to shoot three people to the Moon. He never stopped imagining what it would be like to travel to space, and wrote at length about the required technology for such a feat. One of his ideas, a 'moon car', had a very science-fiction feel to it, and was scientifically feasible. Had the Second World War not intervened, Oberth and others might well have pursued space exploration well before the actual beginning of the Space Age in the late 1950s.

The first planetary probes didn't fly until the late 1950s, when countries around the world participated in the International Geophysical Year. For the US and Soviet Union, the IGY was an opportunity to shoot rockets and instrument packages up for the purposes of exploring near-Earth space and our planet's magnetic fields. However, the Moon was an obvious target; it's close, and, at least in the context of the political environment of the 1950s and 1960s, getting there offered huge rewards in the form of national prestige and international attention. The political gain of being the first to get to our planet's nearest satellite was a powerful incentive.

Of course, the scientific rewards were tantalising as well. At the time of the IGY, planetary scientists had a general idea of how the Moon was formed and what its rocks were made of, and knew that it was an airless desert with no liquid bodies of water. Yet many questions remained. The Moon contains clues to the origin and evolution of our own planet, but the types of *in situ* studies geologists wanted to make was more than a decade away at the time of the IGY.

Racing to the Moon

Successful missions to the Moon were counterbalanced by failures. Pioneer 0, the first US attempt to reach the Moon, was designed by the US Air Force to go into orbit and use a television camera to send back images of the lunar surface. It also had a micrometeorite detector and a magnetometer to measure the lunar magnetic fields. Unfortunately, the first stage of the Thor missile that was lofting it into space exploded seventy-four seconds after launch on 17 August 1958. Undaunted, the US launched Pioneer 1 the following October, followed by Pioneer 2 and Pioneer 3 in November and December. These were also destroyed by launch failures. The Soviet Union sent three versions of their Luna E-1 in September, October and November of the same year. They, too, were all destroyed in launch mishaps – an inauspicious start for both countries.

Eventually, the Soviet Union sent the Luna 1 'Mechta' spacecraft past the Moon at a distance of 9,600 kilometres, on 4 January 1959. It was supposed to slam into the lunar surface, taking data all the way up to the crash. However, due to upper-stage burn problems, it missed its target and instead became the first artificial satellite to orbit the Sun. Like its predecessor, the Sputnik, Luna 1 was a spherical spacecraft with sensors to detect radiation and magnetic fields. It sent back data for about a day and a half before its batteries ran out of power and the transmitter went silent. The Soviets fared better with Luna 2, crash-landing the first spacecraft on the Moon on 13 September 1959. Luna 3 followed with the first images of the far side of the Moon.

The US launched the Pioneer 4 probe on 3 March 1959; it made a flyby of the Moon a day later. This spacecraft was spin-stabilised, and carried instruments to detect radiation and a camera for photography. It was also supposed to make an impact on the lunar surface, but second-stage burn errors en route actually sent it past the Moon, passing within 59,550 kilometres.

Still, it was the only successful US lunar mission from 1958 to 1964, until Ranger 7 was intentionally smashed into the Moon on 31 July 1964, returning images all the way down. Rangers 8 and 9 transmitted thousands of images as they approached and made planned crash-landings on the Moon. The idea behind these missions was to see close-up features on the lunar surface. NASA had its orders from the late President John F. Kennedy to put a man on the Moon by the end of the 1960s. Missions such as Ranger 7 and 8 were vital to the task. People around the world marvelled at the pictures from their cameras.

As the US and Soviet Union raced to gain primacy on the Moon, each country sent ever more sophisticated probes. The Soviets succeeded in the first soft landing on the Moon with Luna 9. It sent back the first images from the Moon's surface in 1966. The US countered with the Surveyor series, which landed one probe safely while crashing another to the surface the same year. The Soviet Luna and US Lunar Orbiters, along with additional Surveyor missions, sent back more and more images of the Moon, and data about the radiation and magnetic field environment in near-lunar space.

During the years NASA sent astronauts to the Moon, the Soviet Union settled for launching more probes to visit the surface, pick up samples and study the lunar environment. The Soviet Lunokhod rovers were the first such machines to robotically wander the surface and perform experiments. The Soviets continued their Luna flybys well into the 1980s.

Continuing Lunar Explorations
Today, China's CNAS agency, ESA, India's ISRO, Japan's JAXA and at least one commercial player are pursuing lunar exploration. The Chinese began their efforts with the Chang'e lunar orbiter on 24 October 2007. The Chang'e 2 followed in 2010, and also accomplished a flyby of asteroid Toutatis. The third in the Chang'e series became the first to settle onto the surface of the Moon. Chang'e 5-T1 orbited the Moon and then returned to Earth in 2014, an equipment-test precursor for future missions. The China National Space Administration has proposed an ambitious programme of further robotic exploration, and has indicated it will also turn its attention toward sending probes and humans to Mars.

ESA sent the SMART 1 orbital probe to take x-ray and infrared imagery of the lunar surface. Launched in 2003, it used solar-electric propulsion to take it to lunar orbit. It arrived in September 2004 and mapped the surface until its planned impact on 3 September 2006.

Japan entered the lunar exploration arena in 1990 with its Hiten Earth-orbiting satellite. It carried a small lunar orbiter called Hagoromo. The lunar mission was deployed primarily to test and verify technologies for future lunar and planetary missions. Upon its release, Hagoromo entered the wrong orbit, but with some advice from NASA's Jet Propulsion Laboratory, the mission planners were able to guide it into a temporary orbit. Eventually, it was moved to a different orbit so that its on-board instrument – the Munich Dust Counter – could gather data about trapped dust and particles in the near-lunar environment. Hiten was guided to a crash-landing on the Moon on 10 April 1993. This mission marked the first time Japan had accomplished a lunar flyby, orbiter and surface impact, making them only the third nation to do so at the time.

The next Japanese lunar effort was SELENE (also known as Kaguya), launched on 14 December 2007. It orbited the Moon for twenty months and was then sent into a controlled crash on 10 June 2009. The scientific objectives were similar to other lunar probes, specifically being to gain insight into the geologic evolution of the Moon from surveying its surface features. In addition, it made very precise measurements of the Moon's gravitational field.

The Indian Space Research Organisation (ISRO) sent its Chandrayaan-1 lunar orbiter to the Moon on 22 October 2008 and mapped the surface with remote sensors until August 2009. It also carried an impactor called the Moon Impact Probe, which separated from the orbiter and smashed into the lunar south pole at Shackleton crater on 14 November 2008. It made India the fourth country to land an object on the Moon.

NASA remains very involved in lunar studies. The Clementine mission (launched in 1994 and developed with the Ballistic Missile Defense Organization) performed imaging in visible, ultraviolet and infrared light. The probe provided evidence for water in the lunar polar craters, and data from its multi-spectral imager (which allowed scientists to determine the mineral makeup of rocks and dust) also point to water *in* the lunar rocks. That discovery would have profound implications for settlements on the Moon, which would need good supplies of water for any future inhabitants.

The Lunar Prospector mission followed in 1998, and did further searches for water ice in the craters at the lunar south pole. It also provided a high-resolution orbital map of the Moon, before its deliberate crash-landing at Shoemaker crater.

The Lunar Reconnaissance Orbiter (LRO) is (as of 2017) in a polar mapping orbit, looking for safe landing sites for the next generation of landers, rovers and human crews. When it was launched in 2009, the US had not sent a lunar mission for a decade. It has since made up for lost time by creating some of the most accurate three-dimensional maps of the Moon's surface. Launched with the LRO was the Lunar Crater Observation and Sensing Satellite (LCROSS). Its task was to send an empty Centaur rocket crashing to the surface and measure any water vapour or hydrogen in the dust plume that billowed out. About four minutes after the first impact, LCROSS itself crashed into the surface, and observers didn't see much evidence of plumes in the post-impact images. However, further analysis of what *was* seen showed evidence of water ice or vapour kicked up by the impacts into portions of the Moon that had not seen sunlight for billions of years.

The GRAIL probes, designed to do high-resolution gravity mapping of the Moon, were both launched by a single Delta II rocket on 10 September 2011. They were in lunar orbit until 17 December 2012, when they impacted the surface in a deliberate crash landing. The Lunar Atmosphere and Dust Environment Explorer (LADEE) studied the lunar exosphere and dust environment from September 2013 until its controlled crash onto the far side of the Moon on 18 April 2014.

All the lunar missions, including the Apollo crew sorties to the Moon's surface, comprise a scientific 'mother lode' of information about our closest satellite in space, and the history of our own planet. Not only have scientists pinpointed just how and why the Moon is such a scarred world, evidence from its rocks points to an active volcanic past. Comparisons of lunar samples with Earth rocks tell a tale of a mighty collision between the young Earth and a Mars-sized impactor some 4.5 billion years ago. The Moon formed from leftovers of that collision. Impacts were a constant source of planetary formation and destruction in the early solar system, sprinkling craters across Mercury, Venus, Earth, Mars and beyond. The kind of science done by lunar probes is invaluable for planning of future missions, whether robotic or populated by human crews.

The Exploration of Mercury and Venus

The two innermost planets of the solar system present very tough challenges for exploration. Yet, missions to those worlds have helped planetary scientists fill the gaps in our overall knowledge of the planets.

Mercury

Mercury orbits the Sun at an average distance of 58 million kilometres and is subject to extremes in heat and solar radiation. Its day (the time it takes to spin once on its axis, 176 Earth days) is nearly twice as long as its eighty-eight-day orbit around the Sun. Surface temperatures range from 427 C to -173 C. The planet spins very slowly on its axis. Mercury has a very, very thin atmosphere, mostly ionised hydrogen and helium provided by the solar wind and traces of oxygen outgassing from the crust.

Several spacecraft have done flyby observations of Mercury, and one achieved a long-term orbit around the planet and then crashed into the surface. The first look at Mercury came from NASA's Mariner 10 mission, launched on 3 November 1973. The spacecraft picked up a 'gravity assist' at Venus (and took data as it swept past), and made three close flybys of Mercury on 29 March 1974, 21 September 1974, and 16 March 1975. Mariner 10 captured images of about 45 per cent of Mercury's surface, giving the first up-close looks at a planet that (from Earth) only appears as a bright dot in the sky after sunset or before sunrise. From the information Mariner 10 provided, however, planetary scientists were able to determine that Mercury has a magnetic field that's about one percent as strong as Earth's magnetic field. The presence of that field means Mercury has some means of generating one from material moving around in its core. The images showed a planet that looks more like a cratered minefield, riddled with surface cracks and ridges.

The only other mission dedicated solely to this planet was the Mercury Surface, Space Environment, Geochemistry and Ranging spacecraft – nicknamed MESSENGER. It left Earth on 3 August 2004 and looped through the inner solar system several times, shedding velocity so that it could enter Mercury orbit. It arrived on 18 March 2011. Once it made orbit, MESSENGER imaged and mapped the surface, measured the tenuous atmosphere, and mapped the magnetic field of the planet. The mission was so successful NASA extended it by a year to get more images and data. The high-resolution images revealed more details of the cratered and cracked terrain, and MESSENGER's atmospheric instruments found evidence of water in the exosphere. It also found evidence of icefields in the shadowed polar craters.

MESSENGER's mission ended when it ran out of manoeuvring propellant. As its orbit decayed, mission planners programmed the probe to take images and data all the way down to the surface. The probe was intentionally crashed on 30 April 2015.

Venus

Venus lies much farther away from the Sun than Mercury, but still presents some tough obstacles to exploration. For one thing, its thick cloud layer prevents mapping missions from seeing the surface in visible light. Anything landing on the surface is subject to sulfuric acid rains in the atmosphere on the way down, plus incredibly heavy atmospheric pressure (ninety-two times that of Earth's) on the surface. Continuing volcanic eruptions and very high temperatures (470 C) make it a most un-Earth-like place. One Venus day is equal to 243 Earth days, and it rotates on its axis from east to west (rather than west to east, as Earth does). Being closer to the Sun than Earth, Venus's year is shorter – 224.7 Earth days, as opposed to our 365 days.

In 1961, astronomers started using Earth-based radar to study the Venus surface and gain some understanding of the planet's rotation period. Radar mapping continued through 1985, but the exploration of Venus by spacecraft revealed this world for what it really is.

The first robotic probe to flyby Venus, in 1961, was the Soviet Venera 1, but radio contact was lost just prior to arrival. However, it did demonstrate what was needed for interplanetary craft: solar panels to generate power for the various instruments, a telecommunications receiver and transmitter, a propellant-fuelled engine to make course corrections as needed, and a three-axis design to spin-stabilise the spacecraft while it is in transit to its target.

NASA achieved a successful flyby of Venus in 1962 with the Mariner 2 spacecraft, which also marked the agency's first successful planetary encounter. Mariner 2's main accomplishment was the discovery of a very weak magnetic field.

The Soviets were the first to send spacecraft into the atmosphere of Venus and eventually land them on the planet. After six failures, they finally succeeded. On 15 December 1970, Venera 7 transmitted data about the atmosphere and surface temperatures for twenty-three minutes before shutting down. Venera 8 was the first successful lander, settling to the surface on 22 July 1972. It was followed by Venera 9 on 22 October 1975, which sent back the first black-and-white images of the surface. They showed a volcanic desert, covered with thick acidic clouds floating in the oppressive carbon dioxide atmosphere. Venera 10 followed up and sent back detailed images of volcanic rock taken by its lander. The first colour images were sent back by Venera 13 and 14 in 1982. The orbiting Venera 15 and 16 spacecraft radar-mapped about 25 per cent of the surface during their missions in 1984–85.

In May 1978, NASA sent the Pioneer Venus 1 mission (also known as Pioneer 12), with seventeen planetary science experiments including cloud measurements, surface mapping, atmospheric studies and gravitational scans among other studies. It orbited the planet, sending back data until 1992.

Pioneer Venus 2 (also known as Pioneer Venus Multiprobe Mission) arrived at the planet on 9 December 1978 and sent six probes into the atmosphere. One of the six actually survived its fall through the thick air and sent back data for about an hour after it landed. Data from these two missions gave planetary scientists a better understanding of the Venusian atmosphere and how the planet interacted with the solar wind.

Planetary scientists took a different approach to Venus exploration with the Magellan mission launched from the space shuttle *Atlantis* in 1994. The spacecraft was originally named the Venus Radar Mapper, and was made up of leftover parts from other missions. It carried high-resolution radar scanners that could operate in three modes: synthetic aperture, altimetry and radiometry modes. It emitted radar pulses to the planetary surface and then used the 'return' signals to map surface features. The mission was very successful, yielding a high-resolution 'look' at Venus's volcanic plains and mountains. It found pancake-shaped volcanic flows, cracked terrain called 'arachnoids' (because some of them look like spiders) and large impact craters. The mission was extended five times before it was sent into the atmosphere to collect data. It's likely the spacecraft was destroyed by friction with the dense blanket of air, but also possible that bits of wreckage survived to crash onto the surface.

At the same time Magellan was mapping Venus, other spacecraft passed by. Typically, they were getting gravity assists on their way to other targets. These included the Vega 1 and Vega 2 probes (sent by the Soviet Union). The pair was originally launched to study Halley's Comet as it rounded the Sun during its perihelion passage in early 1986. The Galileo spacecraft, headed to Jupiter, swept past Venus in 1990. Cassini-Huygens took images and data at Venus on its way to Saturn in 1998, and the MESSENGER mission to Mercury did some studies during its 2006 flyby.

In 2005, ESA launched Venus Express. It used a magnetometer, spectrometer and radio science antennae to probe the atmosphere until December 2014. The mission mapped the temperature range of the southern hemisphere, found a layer of ozone in the upper atmosphere and discovered a cold layer where dry ice may exist high in the upper atmosphere.

The Japanese probe Akatsuki was supposed to go into orbit around Venus on 6 December 2010. Instead, it flew past the planet. Mission controllers scrambled to analyse what went wrong and take steps to correct its flight. Akatsuki finally assumed a stable orbit around Venus in December 2015 with course corrections in early 2016. Its scientific mission involves looking for lightning in the thick clouds, studying atmospheric gases and heat radiation through the atmosphere and taking the temperature of surface rocks. As part of the mission, Japan also launched a solar sail project called Interplanetary Kite-craft Accelerated by Radiation of the Sun (IKAROS). It was deployed on 8 December 2010 and successfully passed by the planet. Future missions to Venus include proposals by Roscosmos, NASA and India's ISRO.

Robotic Mars Quests

Aside from Earth, Mars is the most explored planet in the solar system. To date, that exploration has all been robotic, with orbiters, mappers, landers and rovers sent to the Red Planet. Space enthusiasts have wholly embraced human missions to Mars as a next big step in exploration. Someday people will be on the Red Planet, but for now Mars remains a planet populated only by robots.

The technical aspects of Mars missions sprang from seeds planted by Wernher von Braun in his 1948 book *The Mars Project*. He planned a mission led by humans to the Red Planet, and hoped to see it happen by 1965. Von Braun foresaw a fleet of seven passenger ships plus cargo haulers that would be assembled in

Earth orbit for deployment to Mars. It was a fine skeleton of a plan, but it lacked one essential component: robotic scouting missions to send back images and data about conditions there. Nonetheless, it got people thinking about what missions to Mars would require.

Since the early 1960s there have been more than fifty direct missions to Mars, plus a handful of flybys by craft on their way to other places. Not every one was successful, but Mars continues to be a target. There's a good reason for that – the Red Planet is very likely the nearest thing to an Earth-like planet in the solar system, though it remains a very challenging environment for humans. Even though living on Mars would be difficult, and would require people to work in space suits and live in specialised habitats, it's possible we could someday colonise the planet.

The Early Years of Mars Exploration
Even as the US and Soviet Union raced each other to the Moon in the 1960s, they also competed to see who could get a spacecraft to Mars. The Soviets were first out of the gate with the IM 1 on 10 October 1960, but it never made it to Earth orbit. A string of failures followed that attempt, including the first of its Mars series. NASA countered with Mariner 3 on 5 November 1964, but it too failed at launch. The first to actually reach Mars was the Mariner 4 spacecraft, which flew by the planet on 1415 July 1964. It was also the first mission to return images of the surface of another planet, and sent back data about interplanetary space until 21 December 1967. Mariners 6 and 7 performed successful flyby missions, sending back atmospheric data and images. Mariner 9 arrived on 14 November 1971 and found the planet in the thick of a major dust storm. It had to wait nearly two months to get its first images. The mission provided a year's data about the Martian atmosphere and surface until it was turned off in late 1972. In the meantime, the Soviet Union kept trying, and finally had a partial success with its Mars 2 mission in November 1971. The spacecraft achieved orbit and sent a lander to the surface. Unfortunately it failed, as did a number of other craft in the Mars series.

Vikings to Mars
The first truly long-term missions to the Red Planet were conducted by NASA when the agency landed Viking 1 and Viking 2 on Mars on

20 July 1976 and 3 September 1976 respectively. The Viking 1 lander settled onto the surface at Chryse Planitia in the Martian northern hemisphere. Viking 2 landed at Utopia Planitia. These big, high-profile spacecraft combined an orbiter and lander in each mission. The orbiters took surface images while the landers got to work digging into the rusty-red surface in search for evidence of life, and used their temperature and wind velocity instruments to measure the weather. All the images and data showed that Mars is a dry, dusty desert planet.

The planetary scientists involved with the Viking mission were interested in questions of life on Mars. Images were not encouraging. No lifeforms showed up; no trees or animals or even tiny insects wiggling in the dust. The landers had cameras and instruments to measure the conditions at each site, but they also carried special life science experiments that took soil from the surface and analysed it for traces of life. None were found, although debate continues to this day about ambiguities in the results. More sophisticated experiments in the future may determine if life existed (or exists) below the surface.

Beyond the Vikings
The multi-year Viking missions provided the best *in situ* look at Mars until a fleet of rovers landed, beginning in 1997 with the arrival of the Mars Pathfinder station and deployment of its mobile rover Sojourner. It was followed in 2004 by the arrival of the Mars Exploration Rovers Spirit and Opportunity, the Mars Phoenix lander in 2008 and Mars Science Lander Curiosity in 2012. Opportunity and Curiosity continue to explore the surface, giving planetary scientists highly detailed looks at rocks, dust and Martian landforms in their vicinity.

In addition, a fleet of orbiters continues to map and image the surface. These include NASA's Mars Reconnaissance Orbiter and MAVEN missions, ESA's Mars Express, and ISRO's Mars Orbiter Mission (MOM). Some of the most striking findings have revolved around the existence of water on the Red Planet. Dry riverbeds, floodplains and desiccated lakebeds give clear visual evidence of the past existence of water. The rest of Mars's water history is written inside its rocks. The rovers have used special instruments to grind into cliffs and boulders and studied their mineralogy to find sandstones and deposits of minerals that formed in the presence of water.

Much of the evidence points not to recent epochs of oceans and lakes but to a warmer, wetter Mars billions of years ago, when the planet was younger and more active. Orbital images show tantalizing evidence that water may still flow from seasonally melted permafrost frozen into subsurface soils. Of course, the polar caps are a source of some water, releasing it mainly into the atmosphere, but icy material may also be found in the loose layer of small rocks, dust and sand on the surface (called regolith). The geological systems of Mars – volcanism, tectonism, cratering and weathering – tell of a world that has changed drastically since its birth some 4.5 billion years ago.

What happened to Mars's water and the causes for its loss are the subject of MAVEN's studies, as well as ongoing observations by MOM. That spacecraft, a highly successful technology demonstrator for India, has been studying the upper atmosphere and cataloguing its chemical composition. The MOM team is also working with the NASA MAVEN mission, which arrived at Mars in September 2014 to study the atmosphere in great detail. MAVEN data have shown the thin atmosphere is significantly affected by solar storms (more so than with other planets), and the solar wind has played a big role in stripping away the atmosphere throughout Mars's history.

The treasury of data collected so far from four decades of Mars exploration is under continual study as mission planners ponder landing sites and locations of future science labs on the Red Planet. That same data will be of immense help to human crews as they plot their explorations of Mars as well.

The Outer Solar System

Due to their greater distances, the worlds beyond Mars have taken longer to explore. Not only must spacecraft travel across millions of kilometres of space, but each planet presents its own special hazards. From radiation to incredibly strong gravitational pull to rings of debris and swarms of moons, the planets of the outer solar system present both danger and opportunity.

Pioneers to Space

The first spacecraft to explore the outer solar system were Pioneers 10 and 11. Launched in 1972 and 1973 respectively, they were sent primarily to scout out Jupiter and Saturn, and measure the interplanetary medium along the way. They were part of NASA's 'Planetary Grand Tour' programme, taking advantage of a fortunate alignment of the outer solar system planets. A spacecraft launched at just the right time would use gravitational assists to swing from one planet to the next without requiring them to carry along costly propellants. Each assist would boost the speed of the spacecraft and deflect it toward its next target. The Pioneers were a way to test out the gravity assist approach. They did carry their own propellant supplies, plus radioisotope thermoelectric generators to power the instruments and subsystems. Each spacecraft's science instrument package included a plasma analyser, magnetometer, charged particle detector, cosmic ray telescope, meteoroid detectors, radiation detectors, photometers (light-sensitive instruments) and a photopolarimeter.

Pioneer 10 began its close-approach study of Jupiter in November 1973, with a flyby on 3 December. It gathered data about the planet's atmosphere, gravitational field and magnetic field. Pioneer 10 also provided the first close-up views of Jupiter, created using an instrument called an imaging photopolarimeter. It then swept out on a trajectory headed toward the area of the sky defined by the constellation Taurus, the Bull. The mission was declared at an end in 1997, although it kept transmitting data successfully until 2003, when its last weak signal was heard. It is now well beyond 100 astronomical units from the Sun.

Pioneer 11 did its close flyby of Jupiter on 3 December 1974, exactly a year after its twin passed by, providing detailed images of the tenuous ring system surrounding Jupiter, and sent back the first views of the polar regions. It then got a gravity assist from Jupiter and headed out to its next target. The spacecraft flew past Saturn on 1 September 1979, and was the first probe to visit. It crossed the ring plane and discovered several small moons. After the flyby the spacecraft continued out of the solar system, heading in the direction of the southern hemisphere constellation Scutum.

The Voyager Missions

The dream of a planetary grand tour came true with the Voyager missions. These twin spacecraft, equipped with many of the same types of instruments as the Pioneer missions, plus vidicon tube cameras to take visible-light images. Each spacecraft was roughly the size of a small car. With its 'eyes' on the giant planets, the Voyagers truly opened up these worlds for our view.

Voyager 1 launched on 5 September 1977. It flew past Jupiter in March 1979 and Saturn in November 1980. Mission planners debated about sending it on to Pluto, which would have been 'on the way' out along the probe's path, but they decided instead to route it past Saturn's largest moon, Titan. That allowed them to get images and data about this mysterious moon, but ultimately put the spacecraft on a path up and out of the solar system. Titan turned out to be the right choice, since planetary scientists were able to learn more about its hazy atmosphere. It was enough to programme the upcoming Cassini mission and its Huygens probe for more detailed studies.

Since the Saturn encounter, the Voyager spacecraft has entered its 'secondary' mission, to explore the extreme boundaries of the solar system in an effort to determine where the Sun's influence ends and interstellar space begins. Along the way, it sent back the first-ever family portrait of the solar system, essentially from the outside looking back toward the Sun. It is currently the most distant spacecraft from Earth, traveling at seventeen kilometres per second in the direction of the constellation Ophiuchus, and in 2012 passed through the heliopause – the outer boundary of the Sun's influence. Several of its instrument systems are still operating, and it periodically transmits data back to Earth. Voyager mission operations will likely cease completely by 2030 as the spacecraft's systems age out and stop operating or are turned off.

Voyager 2 was launched on 20 August 1977, nearly two weeks earlier than its twin. Its trajectory was planned so that it would be able to visit not only Jupiter and Saturn (in 1979 and 1981 respectively) but also the ice giants Uranus (on 24 January 1986) and Neptune (on 25 August 1989). The mission followed in the footsteps of its predecessor, returning high-resolution images and data of the planets. In particular, the spacecraft explored the newly discovered rings of Uranus and Neptune, and revealed details about the icy moons of both worlds. Miranda at Uranus appears to be a moon that once broke apart and rebuilt itself, while Triton at Neptune is a frigid world spouting ice volcanoes.

The mission is now in its 'interstellar' phase. The spacecraft continues to study its nearby environment with a reduced suite of instruments, and transmits data about the interplanetary medium, the solar wind and other aspects of the outer solar system. It is now well beyond 100 astronomical units, and making its way out to interstellar space. It's headed in the direction of the constellations Sagittarius and Pavo, and continues to communicate with Earth.

Voyager 1 took one of the most evocative images of the Space Age on 14 February 1990. It's called the 'Pale Blue Dot' picture, and it shows Earth from a distance of 6 billion kilometres. At that distance, Earth was barely a pixel across. The image was taken at the suggestion of astronomer Dr Carl Sagan, who was part of the Voyager mission team. About it, he said in a talk given at Cornell University on 13 October 1994:

> We succeeded in taking that picture, and, if you look at it, you see a dot. That's here. That's home. That's us. On it, everyone you ever heard of, every human being who ever lived, lived out their lives. The aggregate of all our joys and sufferings, thousands of confident religions, ideologies and economic doctrines, every hunter and forager, every hero and coward, every creator and destroyer of civilisations, every king and peasant, every young couple in love, every hopeful child, every mother and father, every inventor and explorer, every teacher of morals, every corrupt politician, every superstar, every supreme leader, every saint and sinner in the history of our species, lived there – on a mote of dust, suspended in a sunbeam.

The pictures Voyager made when it imaged the Pale Blue Dot were combined to make a 'family portrait of the solar system'. The data took more than five and a half hours to get to Earth. Since then, the spacecraft has left the solar system, on its way through interstellar space.

A New Wave of Outer Solar System Exploration

The Voyager and Pioneer missions whetted the scientific appetite with tantalising tastes of outer solar system data. On 4 May 1989, the Galileo spacecraft was launched from space shuttle *Atlantis* on a multi-target mission past Venus, Earth, two asteroids, and eventually to orbit Jupiter for long-term studies. Despite problems with the main antenna, which caused the team to switch transmissions to a low-gain antenna (thus slowing down the data transmissions), the spacecraft's accomplishments were impressive. It made the first-ever asteroid flyby of 951 Gaspra, went on to study asteroid 243 Ida, and made the first discovery of Ida's moon Dactyl. While en route, it observed the twenty-one pieces of Comet Shoemaker-Levy 9 slam into Jupiter in July 1994.

Once at Jupiter, the spacecraft sent a probe into the cloud-tops of the planet and gave scientists the first-ever 'look' at conditions in the upper levels of Jupiter's atmosphere. It studied the tiny volcanic moon Io (which was seen to be actively volcanic by the Voyager missions), and discovered an active magnetic field at the moon Ganymede, and thin atmospheres on each of the four Galilean moons. The Galileo mission ended on 21 September 2003 and the spacecraft plunged into Jupiter's atmosphere during its final orbit.

No other mission was sent directly to Jupiter until the launch of the Juno mission in 2011. Others, such as Cassini and New Horizons, made flybys for gravitational assists, and took the opportunity to do some planetary science along the way.

Juno arrived in 2016 and began studying the system with a suite of instruments (including imaging). Its main scientific objective was to closely measure Jupiter's gravity and magnetic fields, take detailed images and study the atmosphere of the planet. The mission accomplished thirty-seven orbits, tested the largest set of solar arrays ever used on a planetary probe to date, and returned extremely high-resolution close-up images of the planet, focused particularly on its complex cloud systems.

Cassini at Saturn

The long-running Cassini-Huygens mission is best known for its thirteen-year study of the Saturnian system. It dropped the Huygens probe onto the surface of Titan, which supplied the first-ever images of that frozen moon and data about its atmosphere and surface ices. Along the way to Saturn, the spacecraft flew past Earth's Moon, went on to encounter the asteroid 2685 Masursky (named for Voyager mission scientist Hal Masursky), then whipped past Jupiter for a gravity assist.

Cassini's exploration of Saturn revealed new moons, highly detailed views of patterns in the rings, and an aspect of the moon Enceladus briefly hinted at by the Voyager spacecraft images. That tiny moon appears to have a deep saltwater ocean beneath its icy crust, which is split and cracked. Plumes of water ice jet out from the moon and eventually form the planet's E-ring.

As for the planet itself, Cassini revealed constant changes in the turbulent, cloudy atmosphere, and discovered a mysterious polar hexagon in the cloud and intricate waves in the rings of the planet. Its Huygens probe found lakes on Titan and sampled that moon's hydrocarbon-rich atmosphere.

Before it was flown into the Saturnian atmosphere on 15 September 2017, the spacecraft accomplished four missions: its prime one from in 2004 through 2008; the Cassini Equinox mission extension that lasted until October 2010; the follow-up Cassini Solstice mission; and the ending orbits, called the Grand Finale.

Exploring Pluto

The New Horizons mission we read about earlier in the chapter was a record-setter in many ways. It travelled the greatest distance to its target and took the longest time (more than nine years), visited the most distant world and made the fastest-ever lift-off from Earth. It was the first (and will probably be the only) flyby of the dwarf planet Pluto. It also took the longest of any mission to send back its data trove from the flyby – more than a year and a half of transmission from the spacecraft to Earth. The data, which will be analysed for years, reveal Pluto to be a far more complex planet than people expected.

Close-up images of Pluto supplied by the New Horizons imaging system showed many different kinds of landscapes in the frozen nitrogen crust, ranging from mountains and plains to what could be cryovolcanic features venting material from deep inside. It has a thin atmosphere that appears to be replenished by activity on the planet. Its companion world, Charon, appears to be a grayish charcoal hue, with regions coloured a reddish-brown by deposits of material from Pluto. Four other moons complete the Plutonian system, and the spacecraft caught quick glimpses of all of them. New Horizons is on the way to a second encounter in the Kuiper Belt, with a tiny world currently named MU069. It will fly past on 1 January 2019, and then continue on its trajectory out of the solar system.

The observations of Pluto define the 'third zone' of the solar system, the Kuiper Belt and beyond. Worlds 'out there' may well be as geologically active as Pluto, and may represent a whole new regime of icy dwarf planets that will keep scientists busy for decades.

Exploring the 'Leftovers' of the Solar System

The golden age of planetary exploration hasn't been limited to planets and their moons. Missions have also been sent to two categories of objects that contain clues to the conditions and dynamics at play when the solar was forming some 4.5 billion years ago. These are asteroids and comets. Asteroids are rocky bodies that hark back to the age of planetary formation. At that time, asteroid materials coalesced into planetesimals and protoplanets. The largest masses became the planets; the leftovers are the asteroids. Today, they populate the asteroid belt and are also found throughout the solar system, including in near-Earth space.

Comets are icy bodies that were scattered to the outer edges of the solar system. There are countless cometary nuclei in the Kuiper Belt and the very distant Oort Cloud regions that lie beyond the orbit of Neptune. They occasionally get nudged out of their distant homes and rush into headlong orbits around the Sun. As a comet moves through space, it leaves behind bits of ice and dust. That material contains valuable clues to a comet's chemical composition and the conditions it has experienced.

Some missions have flown past asteroids on their way to other targets. However, in the last years of the twentieth century planetary scientists identified comets and asteroids as interesting and important targets for exploration on their own and began planning direct missions to them. These include Deep Space 1, NEAR, Stardust, Genesis, Hayabusa, Rosetta, Dawn, and OSIRIS-REx. These spacecraft have rendezvoused and gathered bits of interplanetary dust from comets and asteroids, or gone into orbit around them.

Dawn is a multi-year project that orbited asteroid Vesta and then went on to settle into orbit around Ceres (both in the asteroid belt). It studied Vesta for fourteen months, gathering data that will help planetary scientists understand the role that asteroids played in the formative epochs of solar system history. The spacecraft then left Vesta and headed for a rendezvous with dwarf planet Ceres in March 2015. Since its arrival there, Dawn has studied and imaged ice volcanoes that appear to be spewing mineral- and salt-rich ices out from beneath the surface. The spacecraft detected evidence of salt-containing compounds as well as ammonia-rich clays.

Exploring a Comet

The Rosetta mission was one of the most daring and exciting missions ever sent to a comet. The spacecraft was launched on 6 August 2014 and flew past Mars and two asteroids before it arrived at its final target, Comet 67P/Churyumov-Gerasimenko. The spacecraft settled into an orbit around 67P and studied the nucleus with high-resolution cameras and spectrometers as it made its closest approach to the Sun. A tiny lander probe called Philae landed on the comet to gather first-hand information about the icy crust. Unfortunately, it bounced several times and settled into the shadow of a cliff. Its power panels couldn't get enough sunlight and the batteries died after a few hours. That limited Philae's scientific return, but it did supply a few images and some dust and ice analysis. Rosetta's images showed good detail of the surface of the comet. It's instruments also showed the comet's water was not quite the same composition as water on Earth. This may mean that comets like 67P were not the sole source of water early in Earth's history. At the end of the mission in September 2016, the Rosetta orbiter was gently nudged to the surface of the comet, where it remains as the nucleus continues on its way out to the most distant part of its orbit around the Sun.

In chapter 9, we will take a closer look at future missions planned to follow up on the decades of planetary exploration already done. What past and current spacecraft have shown us is a remarkably diverse set of worlds in our own corner of the galactic neighbourhood. What planetary scientists have learned will help them understand worlds around other stars – and give deeper insight into our own world and our existence upon it.

8

Astronomy in the Space Age: Taking Steps to the Stars

Telescopes are in some ways like time machines. They reveal
galaxies so far away that their light has taken billions of years
to reach us. We in astronomy have an advantage in studying the
universe in that we can actually see the past.

Sir Martin Rees, Astronomer Royal of Great Britain

From the moment of using great rocket devices, a great new era
will begin in astronomy: the epoch of the more intensive study
of the firmament.

Konstantin Tsiolkovsky

If you wish to make an apple pie from scratch, you must first
invent the universe.

Carl Sagan, *Cosmos*

It may not be the first thing you think of when you look up at
the stars at night and wonder about where they came from and
what their futures are, but we are in fact related to them in a
very physical sense. We are all star stuff. We came from the stars.
Everything we see – every planet, every asteroid, every nebula – it
all originated in the stars. Dr Carl Sagan made that idea famous
in his TV series and books, and it is absolutely true. The 'stuff' of
your flesh and bones, the rocks in our planet, the water we drink,

the air we breathe, all contain elements created by generations of stars. The science of astronomy helps us explore and understand those distant points of light. In a sense, when we use astronomy to comprehend the origin and evolution of the stars and planets and galaxies, we're looking back at our very origins.

We live in an age when movies, TV shows and science fiction routinely have explorers zipping between the stars at astounding speeds, encountering new civilisations and making amazing discoveries. Often enough, those explorers find new understanding, not just of the stars and planets they visit, but of themselves as well.

In one of his earliest tales, called *The Star*, Arthur C. Clarke's protagonist, a Catholic priest, is part of a future expedition to a vast, glowing nebula some 3,000 light-years from Earth. He and the scientists on the crew find a tiny, burnt-out planet in the debris of what turns out to be a stellar explosion. On that planet is a pylon with the sounds, images and literature of a long-gone civilisation destroyed by the supernova that exploded nearby. The priest comes to an incredibly haunting realisation as he and his colleagues study the aftermath of that long-ago catastrophe:

> I know how brilliantly the supernova whose corpse now dwindles behind our speeding ship once shone in terrestrial skies. I know how it must have blazed low in the east before sunrise, like a beacon in that oriental dawn. There can be no reasonable doubt: the ancient mystery is solved at last. [...] What was the need to give these people to the fire, that the symbol of their passing might shine above Bethlehem?

At the time Clarke wrote this moving story, there was a great deal of interest in using astronomy to understand the source of the phenomenon of the so-called Star of Bethlehem. Over the years, people have suggested supernovae and comets as the celestial phenomenon the early scriptural writers described. One by one, each of those ideas has been disproven by careful study of historical accounts, by scientific observation and through astronomical knowledge. Nowadays, the consensus is that this phenomenon was probably a conjunction of planets. The tale was as much Clarke telling us about humanity's changing perception of the universe as it was a description of the celestial event of a supernova explosion. The scientific story is the province of astronomy.

Of course, the first skygazers didn't start doing science right away. They were observers and they charted what they saw, night after night, day after day. Our modern study of astronomy is grounded in their calendar making and rituals.

The first astronomers had only their eyes with which to study the sky, and traced out patterns of stars that we call 'constellations' today. Those patterns were fixed, immutable, but other objects seemed to move against the backdrop of the sky. Early skygazers charted the motions of objects they first referred to as *planetes*, meaning 'wanderers'. Today we know those as the planets. The constellations are merely a trick of perspective from our point of view on Earth. The stars that make them up are scattered randomly through space, some farther away than others. Understanding them had to wait until early in the twentieth century, when scientists such as Cecelia Payne-Gaposchkin, Subramanyan Chandrasekhar, Henry Norris Russell and many others could develop working theories about how stars worked.

The invention of the telescope changed humanity's 'naked-eye' view of the universe to a highly magnified one, and allowed observers to study the stars and planets and other objects in more detail. Until the late 1950s, astronomy observations were confined to ground-based instruments, which limited what astronomers could see. Our planet's atmosphere supports life, but interferes with our view of the cosmos. Of course, clouds are a problem, but even when the skies are clear, the atmosphere blurs the view, making it difficult to pick out individual objects (such as stars in a crowded cluster). It filters or absorbs many forms of light radiating or reflecting from celestial objects.

To get above atmospheric interference, astronomers began launching telescopes to space – at first on short suborbital hops aboard sounding rockets and then into Earth orbit and beyond. That's when the universe really began to reveal its secrets – when we could extend our ability to study the sky out beyond our atmosphere. Mysterious objects suddenly came into view, such as Scorpius X-1 – a neutron star emanating x-rays. It was discovered from an Aerobee sounding rocket in 1962. Astronomers suddenly had a whole new way of learning about the universe.

NASA launched a series of eight Orbiting Solar Observatory satellites, beginning in 1962. Since then, nearly 100 space observatories have been planned, built and sent to space by NASA, ESA, China, Japan and other countries around the world. About twenty are in orbit right now, and more will come online in the coming decades. The age of space-based astronomy is well underway.

The Multi-wavelength Universe

The definition of astronomy is pretty straightforward: the scientific study of celestial objects, space and the universe. Information about distant objects and events in the cosmos comes to us from the light they reflect or emit. Each beam of light is a treasure trove of information waiting to be analysed using the tools of astronomy. Astronomers gather that information using various telescopes and instruments that gather light.

When observing the sky, astronomers look across the light-years (the distance light travels in a year), seeing across space and time. Their telescopes are the ultimate time machines. In reality, any of us who skygaze are also doing the same thing. We look at something in the sky and we see it as it was. The 'image' of the Sun you see is really eight and a half minutes old – the amount of time sunlight takes to travel to us. The view of Mars can be as little as four minutes old or as much as twenty-two minutes, depending on the positions of the planets in their orbits. The Andromeda Galaxy looks to us as it was more than 2.5 million years ago. The most distant objects astronomers can detect appear as they did when the universe was very, very young.

Celestial objects radiate across nearly the entire range of the electromagnetic spectrum, so to get a truly detailed understanding of them it's necessary to look at all the light they emit. Visible light is what we see with our eyes and what most ground-based observatories detect. Visible light can be observed from Earth as well as space, and space-based observatories such as the Hubble Space Telescope have significant visible-light capability. On the electromagnetic spectrum, visible light is roughly in the centre, at wavelengths between 390 and 700 nanometres. (A nanometre is a billionth of a metre.) The study of objects in visible light is referred to as optical astronomy.

The rest of the electromagnetic spectrum is invisible to our eyes, but easily detected with instruments sensitive to non-visible light. Each 'regime' of the electromagnetic spectrum has characteristics that help astronomers identify the objects that emit them, and there is a branch of astronomy for each part of the spectrum.

Generally speaking, highly energetic objects – supernova explosions, black hole jets and hot young stars – emit x-rays, gamma rays, ultraviolet (UV) and radio waves. Infrared (IR) light radiates from 'warm' nebulae, planets and cool stars. The earliest remnants of the Big Bang, the event that started our universe, give off a faint microwave glow across the sky.

The Infrared Universe

Near-infrared light (closest to visible light in the spectrum) can be 'seen' from Earth's surface, but the more extensive infrared spectrum is difficult to observe except from very high-altitude mountain observatories in very dry regions.

Infrared light opens up a hidden universe. There are regions in the sky veiled behind clouds of dust through which visible light cannot pass. However, infrared light can go right through much of that material, which can show us what's 'inside' those dusty cocoons in interstellar space. In addition, energetic objects heat nearby clouds of gas and dust, and cause them to emit infrared radiation. That gives astronomers a way to study the characteristics of those regions. So, an infrared telescope can unveil a star nursery, trace the existence of stars behind or embedded in a cloud and reveal clues to black holes hidden at the centre of our galaxy.

Infrared telescopes also reveal many objects too cool to be seen in visible light, such as aging stars, infrared-emitting galaxies and interstellar molecules. Further out in space, at the earliest epochs of time, very young galaxies emit incredible amounts of energy. As their light traverses space, it gets redshifted – stretched toward longer wavelengths. By studying this stretched light, infrared telescopes function as windows on the most ancient objects in the universe.

Scanning the Ultraviolet Universe

Some of the most energetic objects in the universe give off prodigious amounts of ultraviolet (UV) radiation. Most UV is absorbed by Earth's atmosphere, so astronomers wanting to study the ultraviolet cosmos must do it from very high in the atmosphere or well above it.

Typically, the hotter and more energetic something is, the more UV light it gives off. Massive young stars are a good example. They bathe their surrounding birth clouds of gas and dust in UV, which heats them and causes them to glow in infrared and visible light. It also 'eats away' at the stellar crèches in a process called 'photoevaporation'. So, astronomers can use UV light as a tracer of star formation activity, in the Milky Way and other galaxies. Our Sun emits UV, as do such energetic events as supernova explosions. In addition, gas atoms flowing from comets can become excited (heated) during interactions with the solar wind. That emits UV, which can clue astronomers in to activity in and around the coma. By looking at the UV coming from an object or event, astronomers can deduce its temperature, density, chemical composition, and sometimes even aspects of its magnetic fields. These provide in-depth information that would be missed if we only studied the universe in visible light.

Space-based observatories sensitive to UV have a long history, stretching back to the 1960s. NASA's Orbiting Astronomical Observatories had ultraviolet-sensitive instruments. The first dedicated UV satellite was launched in 1978 – the International Ultraviolet Explorer (IUE), a joint project of NASA, ESA and the UK Science Research Council. It lasted for nearly eighteen years. It was a 45 cm Ritchey-Chretien telescope, outfitted with a beryllium-coated mirror. Light from its targets was directed into spectrographs – instruments that separated the light into a spectrum. The light was then directed to a UV-to-visible converter and then on to a pair of cameras.

IUE provided the first in-depth surveys of the UV universe, from the solar system out to active nuclei of distant galaxies. There have been eighteen other UV telescopes flown in orbit. Of those, NASA's Swift gamma-ray burst telescope (which has UV-sensitive instruments), Japan's Hisaki mission and India's Astrosat are still returning data from Earth orbit. The Hubble Space Telescope has the ultraviolet-sensitive Space Telescope Imaging Spectrograph aboard, and the Solar and Heliospheric Observatory (SOHO, a joint project of ESA and NASA) carries the Extreme Ultraviolet Imaging Telescope to study coronal activity in the Sun's outer atmosphere.

Radio Astronomy

The radio universe is a busy place, with radio-loud objects ranging from the planet Jupiter and our own Sun to distant radio galaxies, quasars, newborn stars, supernova remnants, pulsars and black holes.

The science of radio astronomy depends on antenna dishes to gather radio waves for study. At the present time, only one radio telescope is orbiting Earth – the Spektr-R (RadioAstron) mission deployed by Russia in 2011. Since radio astronomy can be done from the ground, there aren't many orbiting telescopes focused on this part of the electromagnetic spectrum. However, the microwave portion of the spectrum is better observed from space.

The microwave universe really gives astronomers a look at the earliest epochs of cosmic history. Microwave astronomy began in the 1960s when researchers at Bell Labs in the United States detected background noise while testing a special antenna. It wasn't just from one source, but seemed to be coming from all over. After further testing, this noise was named the Cosmic Background Radiation. It fills the universe and is now known to be a faint signal from the Big Bang, which occurred some 13.7 billion years ago.

Charting the X-ray Universe

The x-ray sky is populated by extremely energetic objects with gases heated up to millions of degrees. The Sun is a prodigious x-ray emitter, and in 1949 it was the first object studied with high-altitude sounding rockets equipped with x-ray detectors.

The first celestial x-ray source outside of our solar system was discovered in 1962. It's called Scorpius X-1 and appears to us as the strongest x-ray source in the sky – a binary star system with a neutron star in orbit with a low-mass companion. Periodically, material from the companion gets drawn toward the neutron star, where it falls onto the surface and releases a huge amount of energy, including x-rays.

Other sources of x-rays include the active cores of galaxies (where material cycling into black holes may be superheated as it swirls into the singularity), supernova remnants and hugely massive stars called cataclysmic variables.

The Chandra X-Ray Observatory and XMM-Newton are several of the best-known x-ray detectors of the current fleet, along with China's Hard X-ray Modulation Telescope.

The Extreme High-energy Universe

While x-rays usually have high-photon energies, gamma rays are the most energetic of the electromagnetic spectrum. Gamma rays have been detected by astronomers since the 1960s, and, like x-rays, are largely filtered out by our atmosphere. So, early satellites such as the Orbiting Solar Observatory 3 gave astronomers the chance to look for more of this type of radiation.

Gamma rays are produced by the most violent events and unusual objects in the universe, as well as by our Sun. Some common sources are supernovas (created when massive stars die in cataclysms) and objects such as neutron stars and black holes. However, there are others that astronomers are still working to understand.

In the 1960s the Vela constellation of military satellites began detecting what looked like gamma-ray flashes from nuclear bombs, but they weren't occurring on Earth. They were coming from across incredible distances of space, and got the name gamma-ray bursts. They could be as long as a few minutes or as short as a few fractions of a second, but these events indicated something very energetic was happening. Not only were the explosive events bright in gamma rays, but their afterglows could be seen in visible light.

Something bigger and more energetic than even a supernova had to be causing them. But what? Here's one idea: the merging of binary neutron stars. A neutron star forms when a very massive star collapses in on itself during a supernova explosion. The core of the former star becomes compressed into a very dense object made entirely of neutrons. These are very dense objects, with strong gravitational and magnetic fields. When two of these collide, gamma and other high-energy radiation is emitted in a brilliant, short-lived flash that then fades away.

Particle Physics in Space

The Alpha Magnetic Spectrometer 2 (AMS-2) is installed on the ISS and is used to measure cosmic rays and other energetic particles coming in from space. As of mid-2017, it had detected well over

120 billion particles, using a supercooled superconducting magnet that creates a strong magnetic field to 'capture' particles as they pass through. There are several aims for this experiment: to search for the existence of anti-matter (theoretically predicted to exist), to look for evidence of dark matter (known to exist, but not directly observed) and to measure the amount of cosmic rays in our region of space. While cosmic rays mostly pass through us (and Earth) harmlessly, we are shielded here on our planet. However, exposure to cosmic radiation over a long period in space (such as during a trip to Mars) could pose significant hazards to travellers.

The Great Observatories

Just as no celestial object radiates only one form of light, there is no one truly multi-wavelength observatory that can detect all the light from distant stars, planets, nebulae and galaxies, so astronomers have outfitted telescopes with detectors sensitive to specific ranges of light and energetic particles such as cosmic rays. Some observatories, such as the Hubble Space Telescope, carry instruments sensitive to ultraviolet, visible and infrared wavelengths. Sensors on the XMM-Newton x-ray telescope launched by the ESA, for example, are restricted to much narrower parts of the electromagnetic spectrum.

The Great Observatories programme was proposed in 1979. The idea was to launch and orbit observatories to scan the widest possible range of the electromagnetic spectrum: visible, gamma rays, x-rays, ultraviolet and infrared. Their missions would overlap so astronomers could observe objects at the same time across the spectrum as needed. Eventually, four orbiting telescopes were funded and built for the programme: the Hubble Space Telescope (launched in 1990), the Compton Gamma Ray Observatory (launched in 1991), the Chandra X-Ray Observatory (launched in 1999), and the Spitzer Space Telescope (launched in 2003). Of the four, only the Compton has ceased functioning and was de-orbited in 2000.

Hubble Space Telescope

The Hubble Space Telescope (HST) is the most famous orbiting observatory. It orbits the planet once every ninety-seven minutes, at an altitude of 569 kilometres. It has been in orbit since 26 April 1990, the first of the Great Observatories to be launched.

Currently HST contains the Wide Field Camera 3, the Cosmic Origins Spectrograph (sensitive to UV), the Advanced Camera for Surveys, the Space Telescope Imaging Spectrograph (sensitive to visible, some ultraviolet and near-infrared light), and the Near Infrared Camera and Multi-Object Spectrograph (sensitive to infrared light). Since its launch, the telescope has made well over 1.4 million observations, looking at objects as close as Mars and Venus to objects out at the limits of the observable universe (about 13.4 billion light-years away). The information from its observations is archived at the Space Telescope Science Institute in Baltimore, Maryland, and comprises well over 140 terabytes of observational data.

The HST was built beginning in the late 1970s, but the idea of an orbiting observatory wasn't new. The seeds for Hubble were planted in a book by German rocket engineer Hermann Oberth. In *Die Rakete zu den Planeträumen (The Rocket into Planetary Space)* he described his seemingly fanciful idea about building a telescope in orbit, well beyond Earth's blurring atmosphere. This idea caught the attention of theoretical physicist Lyman Spitzer Jr, who, in 1946, wrote a paper outlining the advantages of having a telescope in space. He and other astronomers wanted to be able to see smaller and dimmer objects in space but had to fight Earth's atmosphere to do so. Ground-based telescopes were limited in resolution (the ability to distinguish individual objects in space); to give the scientific explanation, they were limited to about 0.5 to 1 arcsecond. If a telescope could be launched above Earth's atmosphere, Spitzer (and others) reasoned that it would be possible to achieve higher angular resolution, getting 0.05 arcseconds using a 2.5-metre mirror.

Higher resolution wasn't the only 'Holy Grail' driving astronomers to seek space-based solutions. An orbiting telescope could also detect ultraviolet and infrared light, as well as gamma-ray and x-ray signals from distant objects. So, the astronomy community set to work designing ideas for such an observatory. In the early 1940s, rockets were able to carry instruments to high altitudes, allowing astronomers to study the Sun. In addition to high-altitude sounding rockets, the Orbiting Astronomical Observatory series had already proved that space astronomy could be done.

NASA submitted a proposal for a multi-wavelength space telescope in 1974. The Large Space Telescope (LST) was a complex observatory project with a very high price tag. Both factors sent the scientists back to the drawing boards, and also to Congress to lobby for funding. They had to scale back the size of the proposed mirror from 3 to 2.4 metres, and other design changes were made. Finally, after soliciting funding and an instrument from ESA (to make the project multinational and thus spread out the cost to other countries), Congress approved funding for the LST in 1977. In 1983, it was formally named after Edwin P. Hubble, the astronomer who discovered the expansion of the universe from observations of a variable star in the Andromeda Galaxy.

Construction of the telescope was managed by NASA's Marshall Space Flight Center and the instrument design and build were under the control of the Goddard Space Flight Center, which was also designated as the ground support site. The Hubble Space Telescope Science Institute was built to house telescope data, scientists and mission scheduling and outreach operations. Hubble's main mirror was designed and built at Perkin-Elmer (now Hughes-Danbury Optical Systems). A tiny error was introduced during the polishing process at Perkin-Elmer that wasn't discovered until the telescope was in orbit.

Originally, Hubble Space Telescope was slated for launch aboard a space shuttle orbiter in the late 1980s, but the *Challenger* shuttle disaster pushed back its deployment while the agency investigated the accident. After the orbiters began flying again, HST was scheduled for launch. It lifted off from the Kennedy Space Center aboard space shuttle *Discovery* on mission STS-31, deployed, and readied for use.

Spherical Aberration

Immediately upon receiving the first images from HST, astronomers noticed problems with their focus. At first they thought these would resolve during orbital verification and testing, but that didn't happen. Images continued to look fuzzy and out of focus. Worse, the fine 'collimated' point of light the instruments needed for their work was too spread out. After weeks of testing, Hubble astronomers were faced with the bad news: the telescope had a condition called 'spherical aberration'. The telescope's mirror could not focus light to get the high resolution astronomers wanted. It all stemmed from the tiny grinding error, which ended up with the polished mirror being wrongly ground by an amount smaller than the width of a human hair.

The $1.5 billion observatory immediately came in for searing public attention, bad press and political criticism. It was branded a 'techno-turkey' and NASA and ESA were embarrassed by the attention. However, in the background, scientists were already plugging away on 'work-arounds'. These included special algorithms to 'deconvolve' the images to take into account the effect of spherical aberration on the light coming in to the telescope. NASA granted contracts to Ball Aerospace to devise corrective optics for the instruments. As astronomers worked to eke out the science from the telescope's data, planning went on for the first of five servicing missions to the telescope. It had been built with these refurbishment missions in mind – the idea was to swap out instruments on a regular basis, to replace ageing equipment with newer cameras and spectrographs.

The first servicing mission installed the corrective optics on the telescope. It occurred in December 1993, and replaced the high-speed photometer with Ball's Corrective Optics Space Telescope Axial Replacement. Solar arrays were also replaced, and the mission basically restored the telescope to good working order. The rest of the servicing missions, in February 1997, December 1999, March 2002 and May 2009, regularly repaired ageing electronics, swapped out instruments, and – during the last mission – left a fully functioning telescope in orbit. Hubble should last well into the 2020s.

Hubble's Discoveries

The Hubble Space Telescope was designed and built to give astronomers high-resolution views into the universe. It has been doing so since 1990, enabling discoveries about objects and events that existed and occurred 100 million years after the Big Bang (which occurred some 13.7 billion years ago). The Hubble Deep Field images were the first to show the universe contains galaxies for as far as can be seen in nearly every direction. Astronomers have used its data to more precisely determine the age of the universe, as well as its expansion rate. The telescope has allowed us to peer into the regions surrounding black holes, and to directly detect their effects on surrounding material. The telescope has been used to observe supernova remnants, star clusters, planetary nebulae (the aftermaths of the deaths of stars like our Sun), the births of stars, and most of the planets in our own solar system.

There is enough data flowing from the telescope's instruments to keep another generation of astronomers and their students busy. More than 10,000 papers have been published based on HST results, a wide variety of books have been written, educational videos about its findings are easily found on the web and several planetarium shows have been produced. Art gallery exhibitions of Hubble images bring its discoveries to audiences around the world, and some of its iconic images show up on TV shows and in movies. Even designers have gotten 'Hubble vision', printing up specialty material for garments and home decorations using HST images.

Hubble is not the only multiple-wavelength observatory in orbit these days. In September 2015, it was joined in orbit by the Indian Astrosat, built and launched by the ISRO. It covers an astonishing range of light, from optical, infrared, ultraviolet, low- and high-energy x-ray parts of the electromagnetic spectrum. If it detected gamma rays and high-speed particles, it would very nearly be the perfect multi-wavelength observatory!

The Compton Gamma Ray Observatory

Gamma rays are among the most energetic emissions from active regions in the universe. The Compton Gamma Ray Observatory (CGRO) was launched in 1991 (and de-orbited in 2000) to study the objects and events giving off gamma-rays. It was named for Arthur Holly Compton, one of the earliest experts in high-energy physics, and who won the Nobel Prize for his work. CGRO used specialised instruments to detect high-energy bursts of gamma rays as it performed the first all-sky survey in the gamma-ray spectrum. It found nearly 300 new sources and recorded emissions from almost 3,000 others. Astronomers learned that these gamma-ray bursts (GRBs) exist throughout the universe, in all directions, from objects that lie at great distances from us. They remain something of a mystery, but may be associated with energetic galaxies.

CGRO lasted for nine years before a gyroscope failed (which affected its ability to stay on track while studying sources). Although its instruments were still working perfectly, controllers could not adequately steer the spacecraft so NASA de-orbited the observatory on 4 June 2000. Most of it burned up during re-entry into Earth's atmosphere, but some larger pieces fell into the Pacific Ocean.

A new generation of orbiting telescopes, such as the Fermi Gamma Ray Space Telescope (a joint project of NASA, France, Germany, Italy, Japan, Sweden and the US Department of Energy), continue to gather evidence of gamma rays from the sky.

Chandra's Exploration of the Cosmos

The Chandra X-ray Observatory was launched by NASA from the space shuttle *Columbia* in 1993, the largest telescope ever deployed from an orbiter. It was originally proposed as the Advanced X-ray Astrophysics Facility. In 1998 a worldwide contest to rename the telescope was held, with the winners suggesting the name Chandra, after Indian astrophysicist Subramanyan Chandrasekhar. He is best known for showing there is an upper limit to the mass a white dwarf star can have, now known as the Chandrasekhar Limit. His namesake telescope now orbits in his honour, studying the cosmos from nearly a third of the way to the Moon.

Because x-rays are so highly energetic, they can penetrate an ordinary telescope mirror. To capture them for study, Chandra has an unusual design. Its incredibly smooth mirrors look more like nested glass barrels. The incoming x-rays literally ricochet off the four mirrors and are focused toward the science instruments. Chandra's science instruments include a high-resolution camera, an advanced CCD imaging spectrometer (which allows the scientists to make 'pictures' of objects using only the x-rays emitted by a single chemical element), and two high-resolution spectrometers (which diffract the x-rays and allow for very detailed analysis).

Chandra's Targets

The observations made by the Chandra X-ray Observatory have advanced the science of x-ray astronomy, charting x-ray emissions from the Crab Nebula, the black hole at the centre of our galaxy (called Sagittarius A*), gases spiralling into the heart of the Andromeda galaxy, rings and loops of hot gas around a black hole in a distant galaxy, and odd, newly discovered objects called quark stars. X-rays typically come from superheated material around black holes. Very strong magnetic fields around such objects as neutron stars also play a role in x-ray emissions from these strange, dense objects that formed as massive stars collapsed in on themselves.

Some of Chandra's most recent studies have focused on the growth of black holes in the early universe, and the telescope has been surveying the far distant reaches for x-ray-emitting objects in the young universe. The appearance of black holes early in the history of the universe may tell us something about the distribution of dark matter and the role it played some 13 billion years ago.

Spitzer Space Telescope and the Infrared Universe
As the Large Space Telescope was being planned in the 1970s, astronomers were looking at building the Space Infrared Research Facility (SIRTF). The previous infrared-capable instrument, the Infrared Astronomical Satellite (IRAS), had been sent to space by the US, the UK and the Netherlands, and it provided the first infrared sky survey. The results for deeper studies were very promising, especially for a high-resolution instrument. At first, SIRTF was planned as a telescope that would operate while attached to the shuttle. That plan changed gradually to a self-contained orbiting platform. SIRTF was funded and developed throughout the 1990s, with a plan to put it in an Earth-trailing orbit so the telescope would not be heated by the nearby planet. It was renamed in honour of Lyman Spitzer Jr, the physicist who first began pushing for space telescopes in the 1940s.

The Spitzer Space Telescope has a cryogenic telescope assembly that originally kept things cooled to a temperature just above absolute zero. In order to detect infrared from objects in the universe, the telescope itself must not be a source of heat, which would further confuse observations. On the other hand, the electronics that run the telescope and communicate with Earth have to be kept at near room temperature in order to function.

Spitzer was sent to space atop a Delta II rocket on 25 August 2003. After it went through orbital verification, the telescope began official science operations on 18 December 2003. It carries three instruments: the Infrared Array Camera, sensitive to wavelengths in the near- and mid-infrared range (between 3.6 and 8.0 microns); an infrared spectrograph sensitive to the mid-infrared range (from 5 to 40 microns); and a multiband imaging photometer, a camera sensitive to visible light but which can also 'see' infrared (at 24, 70, and 160 microns). Data are radioed back the Spitzer Science Center at the California Institute of Technology; the mission is managed by NASA's Jet Propulsion Laboratory.

Spitzer's Results

Spitzer was loaded with enough liquid helium coolant to keep its detectors chilled for six years, but that didn't mean the end of the mission in 2009. It merely stopped doing far-infrared observations, and has focused on near- and mid-infrared objects in what is now called the 'Spitzer Warm Mission'. Among its most noteworthy accomplishments, the Spitzer Space Telescope captured the infrared signatures of 'hot Jupiter' exoplanets orbiting other stars. These are planets similar in size and mass to Jupiter, but much warmer. They glow in the light Spitzer is most sensitive to, and astronomers continue to search nearby stars for more such worlds.

Young stars and star nurseries also glow in infrared light. Infrared light can pass through the thick clouds of gas and dust that envelop newborn stars, making Spitzer and other infrared-enabled observatories valuable tools to study these otherwise-hidden phenomena. Astronomers want to know more about the process of star birth, particularly in the very early stages. The general outline is well known – that a cloud of gas and dust somehow begins accreting material in a central region. As the 'core' of this material is pulled together by gravity, temperatures and pressures rise, and the region is threaded by magnetic fields. Eventually, the heat rises enough to create what's called a young stellar object. It's not quite a star, but it's on the way to becoming one. Unfortunately, all this action takes place inside the veiled star birth crèche, and that's exactly what astronomers want to see, using infrared telescopes to peer through the clouds.

Regions where star formation is taking place are found throughout the spiral arms of our galaxy and other galaxies. Star birth also occurs in starburst knots formed in the wake of galaxy collisions. These are all regions Spitzer can study, and its observations are helping to clarify the story of star birth throughout the universe.

Within our own galaxy, Spitzer's observations have focused on the extent of star formation. Spitzer's infrared eyes also allow astronomers see cooler objects in space, like failed stars (brown dwarfs), giant molecular clouds and organic molecules that may hold the secret to life on other planets. When combined with data from other observatories,

Spitzer's observations have contributed to understanding such objects as the Crab Nebula supernova remnant, the filamentary remains of a massive star that exploded thousands of years ago.

Expanding Space-based Astronomy

In the years since the Great Observatories were deployed, astronomers have built ever higher-resolution telescopes with newer technology to improve their view of the cosmos. Each new observatory and the data it returns improves our understanding of the universe.

Herschel Space Observatory

The Herschel Space Observatory began development as the Far Infrared and Sub-millimeter Telescope (FIRST) and was proposed to ESA in 1982. It was built with a 3.5-metre primary mirror and three infrared-sensitive instruments. It was named after Sir William Herschel and his sister Caroline Herschel. Both were astronomers in the eighteenth century, and William discovered infrared radiation as part of his observational research. ESA launched Herschel in 2009.

Like its counterpart Spitzer, Herschel has observed light from the far-infrared to the submillimetre range, giving it a window into some of the cooler and more dusty objects in the universe, including an ocean's worth of cold-water vapour circulating in the accretion disk of a young star. It also found traces of molecular oxygen in space, and studied the water ice streaming from a comet to determine levels of deuterium in the ices. It also discovered filaments of dust and gas that stream through the Milky Way galaxy. Astronomers suspect these filaments are predecessors to the clouds of gas and dust that coalesce to form young stars.

Farther out in the universe, Herschel conducted a 'census' of dust in nearby galaxies to the Milky Way. As nearby stars heat up the dust, it glows in the infrared, making it visible to Herschel's instruments. The telescope operated for four years, running out of coolant in April 2013. Controllers switched off the telescope a few weeks later. Over its lifetime, this observatory made more than 35,000 science observations, returning data that are still being analysed.

Herschel was launched at the same time as ESA's Planck satellite, which was designed to study the universe at microwave and infrared frequencies in order to search out light from the earliest objects in the universe. Planck completed an all-sky survey and its data helped narrow down the age of the universe. In particular, it gave astronomers a high-resolution view of microwave emissions from events and objects that emitted their light only half a million years after the Big Bang.

Insight on the Hard X-ray Universe

On 15 June 2017, the Chinese National Space Agency launched its first ever x-ray telescope, called the Insight (and built as the Hard X-ray Modulation Telescope, or HXMT). Its main targets are black holes, objects with strong magnetic fields (such as neutron stars and pulsars), and the search for gamma-ray bursts. In the field of x-ray astronomy, the so-called 'hard' x-rays occupy the high-energy end of the electromagnetic spectrum. They are second only to gamma rays. Hard x-rays are only detectable from space, and are emitted by thousands of sources. Black holes are among some of the most energetic emitters – although the source of emission is not the black hole itself, but the gas and dust surrounding it. That material is being sucked into the black hole by its intense gravity. As it moves in, the material is accelerated by the gravitational field and heated to very high temperatures, which causes it to emit x-rays. Sometimes material ejected from the black hole's accretion disk is accelerated to very high speeds, and emits x-rays. A more thorough study of these emissions at many black holes will help astronomers trace the action in the vicinity of these otherwise mysterious objects.

Further Exploration of the Microwave Universe

Prior to the Planck mission, which lasted from 2009 to 2013, NASA's Wilkinson Microwave Anisotropy Probe (WMAP) scanned the microwave sky from an orbital point called Lagrange 2 (L2), about 1.5 million kilometres from Earth. It was launched in 2001 and studied the microwave emissions from the early universe, a few hundred thousand years after the Big Bang. What it found was a background glow with a temperature of 2.7 kelvin. WMAP's measurements helped determine the age of the universe to a very high degree of accuracy, but also observed the epoch when the first stars began to shine (perhaps 300 to 400 hundred million years after the Bang Bang) and helped determine the density of atoms in the early universe. It revealed a 'lumpy' universe and generally provided what cosmologists (scientists who study the origins and evolution of the universe) call a 'baby picture' of the cosmos.

The WMAP used a pair of primary reflecting mirrors to scan the sky and focus incoming information onto secondary mirrors, and then to feed horns that focused the data to instruments. The entire WMAP telescope was kept chilled to 90 kelvin. The mission, like many other NASA observatories, far outperformed its 'nominal' parameters, and continued to take measurements until 2010. It was then moved to a 'graveyard' parking orbit.

XMM-Newton

In 1999, ESA launched XMM-Newton, its first major x-ray-sensitive orbiting telescope. As of mid-2017, it's still operational and, Chandra does, gathers x-rays from sources ranging from supernova explosions and active galactic nuclei to energetic emissions from black holes and heated gases in galaxy clusters. It's focused on low- to mid-energy x-rays, and in 2017 ESA published a catalogue of 30,000 objects the telescope has studied. XMM-Newton was one of five observatories to study the Crab Nebula in great detail. A tremendously energetic object, this supernova remnant has been observed by nearly every orbiting and ground-based telescope. At the centre of the Crab is a pulsar – a super-dense neutron star rotating once every thirty-three milliseconds, like a beacon. It gives off emissions across nearly the entire electromagnetic spectrum.

XMM-Newton also uses highly polished mirrors similar to Chandra's to ricochet x-rays through the telescope and onto a focal plane where they are then guided to the science instruments. The satellite is well into its third mission extension and continues to work without problems.

Taking a Stellar Census with Gaia

Astronomers have a good idea of our place in space, and our position in the galaxy. However, measuring and charting the extent of our galaxy, the Milky Way, is a challenging proposition. For one thing, we exist inside the galaxy, and our view towards some parts of it are obscured by clouds of gas and dust. ESA launched the Gaia mission in 2013 to investigate the origin and evolution of our galaxy by conducting a census of at least a billion of the Milky Way's stars. The result will be a three-dimensional map of celestial objects. It will take at least five years to accomplish the charting, and the mission will repeat its survey several times. That will allow astronomers to chart the distance, speed and direction of travel through the galaxy for its target objects.

Gaia has charted the motions of 2 million stars. Scientists have been able to 'plot' their motions into the future to predict what our galaxy will look like 5 million years from now. Galaxies form in mergers of smaller galaxies, and our Milky Way is continuing to gobble up smaller dwarf galaxies. In about 5 billion years it will merge with the Andromeda galaxy, and ultimately will end up as a giant elliptical, perhaps to be called 'Milkdromeda'.

Not only will Gaia be measuring stars, but it will also look for Jupiter-sized planets around other stars. There may be up to 50,000 of these objects in Gaia's census data by the time it finishes the mission. The spacecraft is also an asteroid hunter, spotting the motions of these tiny worldlets in our solar system. Not every asteroid's orbit is known or well calculated, so this data allows planetary scientists to more accurately chart their locations in the solar system.

Kepler's Hunt for Worlds around Other Stars

One of the most evocative and exciting space telescope missions began in 2009 when the *Kepler* mission was launched to hunt for planets around other stars. Named after astronomer Johannes Kepler, the telescope has a 1.4-metre mirror that reflects light into a camera and photometer (a specialised light-meter). The idea behind it is to search for periodic dimming of light from distant stars that could be caused by planets orbiting around them. It measures plots – what are called 'light curves' – which are graphs of the dimming and brightening of the starlight. It was originally pointed in the direction of the constellations Cygnus, Lyra and Draco.

The goal for exoplanet searches is to find Earth-type planets. This means rocky worlds that may have atmospheres and oceans that could support life. Ideally such worlds would orbit in a zone around their stars that would support liquid water on their surfaces. That's the so-called 'Goldilocks Zone' (i.e. not too close, not too far away, not too hot, not too cold, but just right).

Of course, Earth-type worlds would be small and more difficult to spot in the glare of their stars. There are also super-Earths, which are slightly larger than Earth, but not so large as a gas giant-type world. Other types of exoplanets exist, including hot Jupiters, super-Jupiters (just like the name sounds), super-Neptunes and others. These larger worlds block more of the star's light, and also usually orbit farther away from their stars.

The process for discovering exoplanets using Kepler is multi-staged. First, the mission searches for periodic dimmings; intriguing sightings are put on a 'candidate exoplanet' list. Astronomers use that list to plan other observations to confirm the planet, often using ground-based telescopes to zero in on the candidates. In this way, Kepler scientists have identified more than 5,000 candidate exoplanets. More than 2,500 of those have been actually confirmed as planets. Interestingly, twenty-one of those confirmed worlds orbit in the habitable zones of their stars.

The Kepler mission was scheduled to last for three and a half years and was extended to 2016. However, trouble struck the spacecraft in 2012 when part of its pointing system failed. That wasn't the end of the programme, however. NASA accepted a proposal called 'K2', which now allows the spacecraft to be used on a limited basis to continue its search for planets.

Planets aren't the only objects spotted by Kepler. During its K2 mission, the spacecraft has spotted possible binary stars as well as stars that could erupt in certain types of supernova explosions. It is also used to look at variable stars (that is, those stars whose brightness rises and falls due to their own pulsations), flare stars, activity in galaxy cores, even asteroids and comets within our own solar system.

There are other planet searches ongoing, both from the ground and using other spacecraft, so estimates of the numbers of extrasolar planets goes up all the time. Already astronomers have extrapolated some possible planet numbers based on the total numbers of exoplanets found so far. Some astronomers have calculated that there could be at least 50 billion planets in the Milky Way alone. It's possible that some 500 million are in their stars' habitable zones. Most astonishingly, as many as 5.4 per cent of all stars in our galaxy could have Earth-type planets around them. Whether those planets have life on them is an entirely separate question.

To find that out, astronomers need to make careful studies of those worlds, preferably by measuring starlight as it passes through the planet's atmospheres. A signature of oxygen, for example, might indicate the possibility of life. If true, then the astrobiologists would take over, working out just how and why life might be on those worlds, and what it would be like.

Unusual Variations in Tabby's Star
The data from Kepler are available for anyone to comb through and analyse. In 2015, a group of astronomers and citizen scientists published a paper about an unusual star observed by Kepler, called KIC 8462852. It has the nickname Tabby's Star, and is also known as Boyajian's Star. The Kepler data show very large and irregular changes in the star's brightness. It appears in the constellation Cygnus, and has been observed in its brightening and dimming since 1890. The period of dimming happens frequently, with some large dips occurring about once every 750 days. That would argue for some kind of orbiting object or, more likely, a lot of smaller masses orbiting close together.

So, what did Kepler record? One idea is that Tabby's Star could be a young star with a circumstellar disk of material around it. It could be a swarm of comets orbiting together, disintegrating and leaving clouds of ice and dust in their wake. Another astronomer suggested there could be a large ringed planet orbiting the star, accompanied by swarms of asteroids in its wake.

One of the more outlandish suggestions is that Tabby's Star has an artificial structure around it, invoking the Robert Silverberg science fiction book *Ringworld* or perhaps a Dyson sphere – a massive swarm of structures enclosing a star. These, of course, would have been constructed by intelligent life. As with most other ideas like that, scientists suspect that it's not the simplest explanation for what is likely a natural phenomenon. Nonetheless, Tabby's Star continues to brighten and dim, keeping the observational community engaged.

The Future of Space Astronomy

What are the next telescopes astronomers want to use for orbital astronomy? There's at least one mega-project coming on line plus a handful of other long-awaited probes. NASA's James Webb Space Telescope (JWST) was originally proposed as the 'Next-Generation

Space Telescope', and it's the result of a collaborative effort between twenty countries. The main partners are NASA, the ESA and the Canadian Space Agency (CSA). It's designed to enable astronomers to see the universe in visible through mid-infrared light. Its 6.5-metre segmented mirror is designed to be folded up for launch and then 'extended' once the spacecraft arrives at the L2 orbit point. This is shorthand for 'LaGrange point', a stable point in space where a spacecraft can stay in orbit in the same spot with respect to the Sun and Earth.

Just as with other infrared-sensitive observatories, Webb's orbit was selected to keep it well away from heat 'contamination' by Earth. Its planned mission timeline is five years, but could be extended to ten years. Unlike previous infrared instruments and telescopes that needed dewars of coolant to maintain proper temperature, JWST's cooling system consists of sun shields and radiators; proper placement in orbit with regard to the Sun is essential. It carries four sets of instruments and cameras, all designed to give a high-resolution look at the universe out as far as it can see.

JWST will look at many of the same objects in the universe as Spitzer and other infrared observatories. Of particular interest to the science teams is the chance to search out 'first light', the earliest propagation of light that ended the dark ages after the Big Bang. This time is called the 'Epoch of Reionization', when the infant universe had cooled enough to allow light to travel. That was when the first stars began to shine and the first galaxies were forming. The assembly of galaxies since the Big Bang is another key observational aim for the telescope, as well as the births of stars and planetary systems. All these events and objects are very easily visible in infrared and visible light.

While JWST is set to become the eventual replacement for the aging Hubble Space Telescope, it does not have an identical set of instruments, and 'sees' in different parts of the electromagnetic spectrum. With any luck, both will be in orbit at the same time for an extended period, to complement and coordinate observations.

WFIRST

After JWST, the next infrared-enabled observatory planned is the Wide Field Infrared Survey Telescope (WFIRST). It is intended to make observations to expand our understanding of the mysterious 'force', called dark energy, that seems to be affecting the expansion of the universe. It will also search out exoplanets, with assistance from 'microlensing', which is a form of gravitational lensing.

Gravitational lensing occurs when a massive object – say a galaxy cluster – passes between us and a more distant object. The light from the background object is 'bent' around by the foreground galaxy cluster because of its gravitational influence. That makes the galaxy cluster a sort of 'magnifying glass'. This effect is very pronounced when the background object is a bright quasar and its light passes near or through a galaxy cluster. The result is a set of arc-like smeared images of the background quasar. HST has imaged many such gravitational lenses.

If the lensing object is a star with planets around it, then the gravity of those planets can lens the starlight, and that gives away their presence. This technique should work with Earth-size planets and eventually WFIRST's search may reveal how many such worlds exist in our own galaxy. The telescope is planned for launch in the mid-2020s.

Euclid

Dark energy and dark matter are the focus of an ESA mission called Euclid (named for the 'Father of Geometry', Greek mathematician Euclid of Alexandria), still in construction. The idea behind this observatory's design is to measure galaxies and their redshifts (the velocity of their motion away from us) out to a distance of 10 billion light-years, meaning to a point 10 billion years in the past. From that data, astronomers will be able to derive information about the expansion rate of the universe, and how dark matter and dark energy affect it. Both of these remain mysterious because astronomers don't know quite what they are, even though we can measure their effects on the rest of the matter in the universe. In the process, Euclid will add an incredibly useful dataset about billions of stars and galaxies to improve the world's astronomical databases.

TESS

The search for exoplanets is ongoing and the Transiting Exoplanet Survey Satellite (TESS) will use the same technique as Kepler to look for additional worlds around other stars. TESS is still on the drawing boards, with partners at NASA, Massachusetts Institute of Technology (MIT), the Harvard-Smithsonian Center for Astrophysics and the Space Telescope Science Institute participating in the design process.

While Kepler has concentrated only on specific regions of the sky, TESS will survey the entire sky – searching out around 200,000 stars. In particular, the spacecraft will look in detail at stars in the 'close' neighbourhood of the Sun, and observe the light that passes through their planets' atmospheres. This may provide detailed insight into the gases present, indicating whether or not the planets have life.

SPICA

Late in the 2020s, Japan's JAXA agency plans to launch the Space Infrared Telescope for Cosmology and Astrophysics (SPICA). This observatory will orbit at L2 like JWST, but use mechanical cryo-coolers to keep the instruments at the proper temperature for infrared astronomy. Its wavelength range is planned to extend beyond JWST's. The main science objectives for SPICA are pretty basic: how was the universe born? How has it evolved since that time? How are stars and planets formed? What is the history of matter? To answer those, the telescope will study the birth and evolution of galaxies, the processes of star birth and planetary formation.

From Earth to the Cosmos

It has been nearly a century since Hermann Oberth and others began dreaming of telescopes beyond Earth's surface exploring the depths of space and time. It took nearly half that time to actually build and launch the many space-based observatories that have flown. In the decades since, astronomers have used them to expand our knowledge of the cosmos, from planets to the most distant galaxies. Yet that 'knowledge' has come in many forms, beyond simply gathering data and statistics. This kind of exploration has generated many new and unexpected discoveries, harnessing new connections between the many branches of science. The future of space astronomy will give more insight into the fundamental nature of our universe.

9

Next Steps: Where Do We Go from Here?

When a thing has been done, it always looks easy. The years of effort, the mistakes and failures, the arguments with the experts who cried 'Impossible!' are all forgotten. Instead, everyone asks, 'Well, why did it take so long?
Arthur C. Clarke, *Man and Space*

Astronauts will remain the explorers, the pioneers – the first to go back to the moon and on to Mars. But I think it's really important to make space space available to as many people as we can. It's going to be a while before we can launch people for less than $20 million a ticket. But that day is coming.
Sally Ride, interview in *Popular Mechanics*, 2009

Space exploration is a force of nature unto itself that no other force in society can rival. Not only does that get people interested in sciences and all the related fields, [but] it transforms the culture into one that values science and technology, and that's the culture that innovates.
Neil deGrasse Tyson, *Space Chronicles: Why Exploring Space Still Matters*

Questions for the Future: Politics, Private Enterprise and Space Exploration

The future of space exploration seems to be a bright one. More countries are involved than ever before, and private industry is seeing it as not just a way to create a more robust economy but an investment opportunity. That could mean a very different future in space compared to our past explorations. In a 2015 interview, astrophysicist and science promoter Neil deGrasse Tyson offered a viewpoint that may serve as a wake-up call for those who assume the business of space is business. He said:

> Private enterprise will never lead a space frontier. In all the history of human conduct, it's as clear to me as day follows night that private enterprise won't do that, because it's expensive. It's dangerous. You have uncertainty and risks, because you're dealing with things that haven't been done before. That's what it means to be on a frontier … The government is better suited to these kinds of investments. They have a longer time horizon. They're not shackled to quarterly reports like you see in a private enterprise.

Tyson's observations are certainly valid, but the day of only governments handling space activities may have passed. The sheer cost and extent of upcoming projects practically mandates both government *and* private enterprise be involved in future missions. Of course, the role of government may well be to mitigate the 'bottom line' tendency that he is hinting at. That would involve a government ratifying the proper legislation governing such activities. The rhetoric surrounding mining efforts, for example, touts the research opportunities that will be made available. Research for pure understanding of nature (such as that aimed at testing asteroid material to determine the nature of the cloud of gas and dust that formed our solar system) does not always move in parallel with corporate bottom-line goals. It's understandable, since a company is in business to make money. Anything that keeps that from happening isn't going to be a priority. It may well be that good science research *can* be done while helping a company meet bottom-line goals, but we won't know until the first mining labs are set up on an asteroid or the Moon. One only needs to look at corporate science research today on such items as pharmaceuticals and new foods to see some possible parallels.

Other concerns about private enterprise in space involve safety, indigenous life on other worlds and citizenship. Safety concerns in space have always been paramount in government-funded missions. Will this change for private industry? Will space tourists or asteroid miners be subjected to different levels of safety requirements? What will a company's liability be toward its workers and guests? What are the insurance requirements going to be, and how will they change for such high-risk activities as a flight around the Moon? Given the many different standards that exist for such issues here on Earth, will they apply differently in space depending on who you travel or work with? In the United States, worker safety laws are frequently criticised by private industry as 'too restrictive' or 'unfair'. Can a company loosen its rules once it's operating off-planet? Or will the rules in the countries where they operate apply in space, too? For the most part the answers may be found within the language of laws that already exist, and in treaties already signed; however, there are still issues to be addressed. For example, the current interest in mining on the Moon raises questions about the legality of one country (or company) strip-mining resources for the benefit of a few.

Extended exploration on the Moon or Mars also raises the spectre of biological contamination of worlds we have yet to fully explore. For scientists interested in answering questions about life's origins elsewhere in the solar system, the inadvertent introduction of Earth life to what are considered 'pristine' environments would destroy any chance of determining if those places could have (or do) support indigenous lifeforms.

Questions regarding citizenship are raised from time to time. While a Chinese lunar village might 'belong' to China, will the country's laws apply there? Will they affect European nationals living and working on that base? What about Americans or Africans or others who come to visit? It gets even more confusing when you consider a Mars expedition. If the members end up living there for a long time, they will rightly consider themselves independent of Earth. What rules apply then? Who decides? Certainly there are laws we use here on Earth to help govern ourselves, but who says that they will apply to a civilisation that lives in a sealed environment on a planet where people can never go outside without protection? What about people who might live and work in an orbiting colony built by a corporation? Whose rules apply then? The home country of the company (or countries, if it was multinational)?

What about safety in space? Would the Safety of Life at Sea (SOLAS) rules that apply on both military and civilian ships here on Earth be a practical guideline? For that matter, would all the laws that apply to ships be equally applicable to a spaceship? Would the commander of a spacecraft be the de facto ruler, as equally in charge as the master of a seagoing ship?

Because humans are a warlike species, despite centuries of attempts to civilise ourselves, the prospects of war in space are considerable. Treaties to prevent the use of nuclear weapons in space may be worthwhile, but as we have seen in recent decades, wars don't have to be waged with guns and bombs. Information systems can be disrupted by dedicated hackers and will affect space systems, too. Attacks on one country's satellite systems by an aggressor are a real possibility. Economic sabotage is not inconceivable. The race to space could see companies and institutions 'cutting a few corners' to gain economic advantage in, say, resource extraction. It's happened throughout history here on Earth.

Economics is another aspect of space colonisation that raises many questions. It will be interesting to see what type of currency exchange evolves in space. The joint Chinese–ESA lunar village could be an interesting test case economically – the two groups are pooling their resources to plan and build it. What will people on the Moon use for money? Will they need it? Will they be willing to depend on electronic funds or something like Bitcoin? What will they buy? Will companies ship products to the Moon for sale to the colonists once the population gets big enough? How do the colonists get paid? What will be their products? Equally, what will the people on Mars do for money? Imports? Exports?

In the case of a Mars population, it's very clear that these people will be dependent on Earth for only a very short time. The sheer distance from Earth to Mars and back will make it very expensive to send products back and forth. If some new product from Mars did become available to Earthlings, it would be incredibly expensive to ship to a buyer. It's very likely that the Martian population would have to be self-sufficient from the very beginning. On the other hand, lunar goods and services are only a couple of days away from Earth, and possibly could be less expensive to deliver. Of course, opening trade with Earth will require extensive negotiations and may involve tariffs, taxes and import costs that will be borne by those doing the trading. However, such trade will (or should) benefit both sides, and contribute to the creation of thriving economies wherever humans go. That's the upside.

There are darker sides of human exploration, and they're not exclusive to space exploration. Earth's history can show us many times and places where trade only benefitted one side, where exploitation led to conflict. Science fiction stories about 'evil corporate interests' make for good drama, but they're not so far-fetched. Business is business, and someone will always find a way to take advantage of a situation for their own benefit.

Yet we have to hope the interactions between Earth and the humans who migrate to other worlds or orbiting colonies would not reflect the dark parts of our past but the better parts of our human nature. In the future, people should look forward to a peaceful exploitation of space, a chance to enhance our knowledge of the universe regardless of how we enrich some pocketbooks. Whether that happens is up to the world's leaders to work out – perhaps with a nudge from the populations of their countries.

In the not too distant future, a family member or friend may give you a call, or an instant message to tell you, 'I'm going to the Moon!' Or it might be you, making that call to announce your new job in space mining. Going to space is a trip many people have wanted to make since the heady early days of the Space Race, when the US and Soviet Union were doing their best to beat each other to space.

Throughout this book, we've examined the history of space exploration to date, plus taken a look at the agencies and companies involved in taking people and goods to space. The infrastructure is there. Our spacefaring species has the tools to explore, and once we decide to take the steps needed, there will be no stopping those who want to pursue their dreams. The US and the Soviet Union led the first steps off Earth, and they still explore today. China is making big moves towards space, and they are very serious about their work. It could be a lunar village or a trip to Mars or the first asteroid mine.

The country has determination and technical savvy, and their space programme bears watching.

Another country to watch is India. The science community there has made a major commitment to space astronomy, planetary science and eventual human missions. They've staked the future of their country on it, with all the financial and political implications it entails. Both China and India could provide 'Sputnik moments' for the other spacefaring powers in the very near future. Beyond the considerations of starting another space race to the Moon and beyond, however, there are other missions humanity may undertake that will shape our future.

Planetary Defence

Each day Earth is bombarded by tons of 'space dust', much of it from comets and asteroids that intersect our planet's orbit. We see this space debris as it falls to Earth, mostly in the form of meteors that flare across the sky and vaporise in the atmosphere. A larger piece will occasionally fall to the ground as a meteorite, usually causing little damage. However, much larger pieces pose a distinct danger to the planet. Most people have seen videos of the Chelyabinsk impactor that flared across the skies of central Russia on 15 February 2013. It exploded in the atmosphere, spreading heat and shock waves through the nearby area, and eventually pieces slammed into the ground. The worst damage was from broken windows and injuries from flying glass. Experts later calculated that the superbolide (extremely bright meteor) marshalled an energy release equivalent to a 400–500 kt nuclear weapon. Had the impactor landed in a city, the damage would have been considerably more than what the people of Chelyabinsk suffered.

Earth has been bombarded throughout its history; bombardment and accretion is how it was built. Large impactors have periodically collided with the planet, posing significant threats to life. One of the most spectacular impacts occurred some 65 million years ago, when an asteroid chunk about 170 kilometres across ploughed into a location we know today as the Yucatán peninsula of Mexico. It created a huge crater (known as the Chixculub crater) and sent shock waves through the ground and ocean. The material it 'excavated' (called 'ejecta') was lofted into the air, resulting in a global 'winter'. It blocked sunlight, which affected the food chain, and led to the deaths of countless animals and plants. The debris rained down and formed a worldwide layer rich in iridium (a component of the type of asteroid thought to have hit the planet). The dinosaurs began to die out at around this time (and palaeontologists know this based on extensive studies of the fossil records). Their demise is tied to the event, and it marks the end of the Cretaceous geologic period.

Today, planetary scientists study core samples drilled into the impact crater, and using knowledge of asteroid 'families' are narrowing down the possible origins for the impactor. Asteroid families are groups of space 'rocks' that orbit together and tend to have common mineral characteristics. When a piece of one falls to Earth, it provides more insight into the types of asteroids that exist in the solar system.

The study of asteroids encompasses not just the census of families that exist, but also their origins and chemical make-up. In recent years, astronomers have been searching out what are called near-Earth objects (NEOs). These are known to have orbits that come near to or cross Earth's path around the Sun. Each crossing represents a point of possible collision. A census of asteroids reveals where most of the larger ones orbit (those larger than a kilometre in radius). The large ones are easier to spot, but the smaller ones are very challenging. This is not just due to their size; asteroids are not usually bright, shiny objects; they're often dark grey or brown.

Many small asteroids remain to be discovered, and every few weeks we hear of one coming close to Earth. Sometimes they're out at the distance of the Moon; other times they pass by beneath some of our orbiting satellites. We are lucky not to have been struck very often.

The damage from an incoming impactor can be quite extensive, if the Chixculub impactor is any indication. Smaller impacts since then, such as the 1.6-kilometre-wide Arizona Meteor Crater (created by an asteroid chunk the size of a bus some 50,000 years ago) or a pair of craters in Australia believed to have been dug out by two 10-kilometre-wide asteroids, are a testament to the power of something slamming into Earth's surface. At the very least, they dig out extensive holes in the ground and shower the surrounding landscapes with chunks of rock.

The power of asteroids to do serious damage to a civilisation is not out of the question. A small (500-metre) rock smashing into Earth could wipe out a region the size of central London or

Manhattan, depending on its angle of attack and velocity. Even a smaller impactor (say around 300 metres across) could cause a tsunami (if it landed near a coastal region), or wipe out a town.

Detection of Potentially Hazardous Asteroids

While Earth faced a higher chance of impacts in the very distant past due to higher numbers of impactors in our solar system, the chances today are not zero. Earth hasn't swept up all the rocks in its orbit, and they keep wandering in from other parts of the solar system. The space agencies of the world, led by NASA and ESA, are taking particular interest in detecting the where the potentially hazardous asteroids are before they become a threat. Surveys of near-Earth objects show there are more than 600 possible 'high risk' asteroids. That's out of more than 16,191 known near-Earth asteroids out there.

Most of the surveys are made with ground-based telescopes, but some potential impactors have been found using orbital telescopes. The infrared-sensitive Wide-field Survey Explorer (WISE) found more than 33,500 new asteroids and comets as part of its survey work searching the sky. It finished its primary mission, was turned off and then recommissioned for seeking out asteroids. It is now called NEOWISE and continues its search for objects that could pose a threat to Earth as well as smaller asteroids in the solar system.

NASA has implemented a near-Earth object (NEO) Observations programme to find, track and characterise NEOs (which includes both near-Earth asteroids and comets), and in particular, chart those that can come within 50 million kilometres of Earth's orbit. Planetary scientists at universities and observatories, and even amateur observers, participate in the search. Ultimately, the data archive is used to chart the trajectories of these objects. There may be as many as 25,000 of them out there. About 95 per cent of NEOs larger than a kilometre in radius have been catalogued and more are found each year.

Our ground- and space-based technologies for detecting these objects comprise an important part of human space exploration. But what do we do if we find an incoming impactor?

Science fiction movies have long played with scenarios for heroes to go out and 'save the world', literally. The reality is that orbital mechanics and physics make the interception and destruction of incoming space rocks a big problem. Do it right and the world is indeed saved. Do it wrong and the problem gets much, much worse. To that end, organisations such as the Secure World Foundation and the B612 Foundation are researching the best ways to use our space technology in the event of a possible impact. In particular, the B612 Foundation (named after the asteroid in Antoine de Saint-Exupery's novel *The Little Prince*) has as its sole aim to protect our planet from asteroid impacts. It has proposed the Sentinel Space Telescope, an orbiting observatory dedicated to searching out asteroids and mapping their trajectories. Knowing the path of an object gives us time to plan in the event that it strays too close to Earth (or is aimed directly at us).

In addition to Sentinel, which is still in the planning phases, the ground-based Large-Scale Synoptic Survey telescope in Chile (coming online in the mid-2020s) will take an active role in the search for incoming space rocks.

Deflect or Destroy?

How can we use space-based resources when a big impactor threatens the planet? There are several methods for impact avoidance. These threat-reduction techniques involve kinetic impactors (which would push the asteroid off its trajectory and into a safer one), gravity tractors (which would require equipment attached to the asteroid and gravity to gently nudge it off track) or explosive-blast deflection. Study groups in the US and Europe have already examined each of these methods. The Europeans formed an international consortium of researchers to implement a project called NEOShield that examines the ways impacts can be avoided. One proposal, from the Russians, proposes using a nuclear warhead to blast apart an asteroid. A long-standing law against exploding nuclear warheads in space would have to be set aside in the event of what could be a world-shattering catastrophe.

However, each of these methods comes with some problems. First, none of them have actually been developed. Second, they require technology that may be years away from being a reality. In the case of the Russian proposal, yes, the world has nuclear weapons, but how to get one to an incoming asteroid needs to be figured out. Assuming that's solved, then there are other issues. For example, the blast might divert the object into a more dangerous orbit. Blowing it up could then send countless smaller impactors toward our planet instead of moving one big one out of the way, making the situation infinitely worse.

However, we need to have plenty of advance warning to implement any of these solutions. For most known trajectories that warning would be available. However, there is still the problem of asteroids that haven't been detected taking us by surprise. The world has agreed that tracking incoming dangers is important, and our space-based technologies can be harnessed to help solve this – given the right amount time, money and people willing to do it.

Space Weather Hazards

We live near an active variable star called the Sun. Its activity levels rise and fall in cycles of eleven and twenty-two years. When it undergoes strong outbursts, the effects can extend across the solar system, creating a phenomenon called 'space weather'. At its most benign, space weather creates shimmering displays of light called aurorae – the northern and southern lights.

For Earth, space weather occurs when the solar wind, and outbursts from the Sun called flares and coronal mass ejections, collide with our planet's magnetosphere. The magnetosphere is the region around an object where its magnetic field controls incoming charged particles. Earth's magnetosphere shields the planet from the particle-laden solar wind.

Stronger bouts of space weather can damage people and technology. Humans in space are subject to a higher level of ionised particle bombardment from the Sun, which can threaten their health. Astronauts aboard the ISS do have safe areas that are better shielded than other parts of the station, but even so they are at higher risk for cancer and other radiation-induced problems as the problem is worse outside of Earth's magnetosphere. Future missions to the Moon or Mars will be exposed to much more radiation from charged particles originating in solar outbursts.

On Earth, the effects of space weather can be considerable. Propagation of radio signals (including those from GPS) are disrupted by space weather events. Commercial airliners flying through the stratosphere subject the people inside to increased radiation danger at periods of high solar activity. During extremely strong space weather events, electrical currents high in our atmosphere can disrupt power grids below. In 1989, a strong bout of space weather interfered with the power plants in Canada, tripping transformers and circuit breakers, which shut down electricity and lights for millions of people. Some satellites, particularly those in polar orbits, were affected. Weather satellite communications were shut down, and the space shuttle *Discovery* reported problems with some of its sensors.

Solar-observing satellites such as the Solar Terrestrial Relations Observatory (STEREO), the Solar and Heliospheric Observatory (SOHO) and the Solar Dynamics Observatory (SDO), along with Earth-based solar-observing facilities provide constant data about solar activity. The US National Atmospheric and Oceanic Administration launched the Deep Space Climate Observatory to keep watch over terrestrial weather as well as disturbances caused by increased solar activity. The data from the world's land and space-based observatories feeds into such facilities as the Space Environment Lab in Boulder, Colorado, which supplies warnings to commercial and government clients. In particular, satellite operators are concerned about space weather's tendency to affect our atmosphere, which can increase drag on the satellites. Disrupted GPS signals affect navigation for planes, trains, barges, ships and cars, and in a very strong event can also disrupt timing signals needed for international commerce, cellular networks and satellite TV and radio. Our technological society remains at risk due to the vagaries of space weather, and our space exploration assets provide tools scientists use to understand and mitigate those risks.

Mitigating the Space Junk Problem

Ever since the dawn of the Space Age, humans have been littering near-Earth space. More than 800 million pieces of space 'junk' orbit the planet, traveling fast enough to cause major damage to spacecraft and satellites if they happen to collide. The pieces range in size from paint chips to cameras to pieces of satellites blasted apart by collisions with each other. Those debris collisions often produce more space junk, which makes the problem worse. A defunct Russian satellite orbited out of control in 2009 and eventually collided with an Iridium satellite, which added more than 2,000 pieces of debris to the growing collection. In 2007, China performed an anti-satellite test, using a missile to destroy an older weather satellite. That sent more than 3,000 chunks of debris spinning through low Earth orbit.

Every launch faces a small chance of running into orbiting debris. Some of the pieces are large enough to be tracked by the US Strategic Air Command and North American Air Defense Command, for example, and they help launch planners work around them. However, smaller pieces aren't necessarily in their sights.

This space junk threatens future access to space. It can and will be limited by the ever-growing amount of orbital debris. However, not only does all the space junk threaten our orbital assets and launch planning, but falling debris can pose a threat, too. Much of it does get vaporised as it falls through the atmosphere, but not all. Larger chunks are often found on land, although most hits the oceans and sinks to the sea floor.

As with incoming asteroids, there are only a few alternatives to fix the problem of space junk. Shooting the larger pieces down risks making more of them. Going to get the debris would guarantee its removal, but it's an expensive proposition. There are proposals for 'dragnets' to scoop up the larger pieces, and to figure out ways to de-orbit larger ones safely over water. The best course of action is stop junking up low Earth orbit, and space agencies have devised 'best practices' for mitigating the growth of space junk.

Looking at Future Human Missions

As we look to the future of space exploration from our vantage point in the second decade of the twenty-first century, exciting opportunities await. We've looked at astronomy and planetary missions for the near term in chapters 7 and 8, but what's on the far horizon for solar system exploration? What new human missions will take place? It's worth taking a look at some future plans.

Going to the Moon

The Moon is such a close and obvious target for near-future human missions. The Chinese space agency, CNAS, has very ambitious plans for lunar exploration, which started in 2004 with the Chang'e robotic probes, which performed mapping and imaging using orbiters, landers and a rover. These were all in preparation for a 2018 sample-return mission. In 2016, the CNSA announced a joint plan with ESA to build a lunar base within the next two decades. This 'Moon Village' concept is intended as a jumping-off point for other kinds of missions, including planetary exploration. Mining is one possible activity, as well as tourism.

The US is looking moonward, with plans to use NASA's Space Launch System and the crew capsule Orion to return astronauts to lunar orbit, perhaps in time for the fiftieth anniversary of the Apollo 8 mission that had astronauts circling the Moon in 1968. In addition, the US and Russia are exploring a cooperative venture to build the Deep-Space Gateway sometime in the early 2020s. It would be a modular habitat circling the Moon.

Building infrastructure on the Moon, regardless of who does it, will be challenging. One idea floated by ESA planners comes straight from a piece of equipment many have seen on Earth: the 3D printer. Such equipment is already in use on the ISS to make tools, and on Earth to make whole structures (rocket engines, for example by RocketLab), so it shouldn't be difficult to do the same thing on the lunar surface. The idea would be to utilise *in situ* resources to quickly build habitats and research areas for teams, who could then proceed more quickly with their work. Other possibilities for habitats include inflatable ones (presumably from Bigelow Aerospace), or pre-assembled modules sent ahead of time for lunar explorers to use as soon as they arrive.

Astronomy on the Moon

The Moon has a unique environment and astronomers have long talked about building observatories there. Its lack of air allows parts of the electromagnetic spectrum to be 'seen' without the atmospheric interference that hampers ground-based observatories on Earth. The lunar far side is also shielded from a source of pollution – the radio 'noise' from Earth. That makes it a great place to put radio telescopes. The Chang'e 3 lander performs ultraviolet astronomy from the Moon, and astronomers would like to have other low-frequency detector arrays in a massive 'radio free' zone. Those might have to wait until lunar villages and research stations are built there, to provide caretakers to service the systems.

Another mission in development is the Dark Ages Radio Explorer (DARE). It would orbit the Moon permanently, performing astrophysical observations in the low-frequency radio range, between 40 and 120 MHz. Emissions from the early universe can be detected within this range, and can tell us a lot about conditions at a time when the first stars and galaxies were just starting to form. Capturing the light of those earliest objects has long been a Holy Grail for cosmologists – astronomers who study the origin and evolution of the universe.

Detecting Gravitational Waves from Space
The detections of gravitational waves from Earth have opened up new areas of study for astronomers. These waves are created when black holes or massive neutron stars crash together in the distant universe and send ripples through space-time. Detecting them from Earth has been done, but it's a very complex and delicate process. What if you could detect these waves from space?

That's the question ESA astronomers asked when they began planning for an audacious gravitational wave detector called LISA, which stands for Laser Interferometer Space Antenna. A 'test mission' was launched in December 2015 and has been trying out detection strategies using test masses as stand-ins for the sought-after black hole mergers. A proposal for the LISA observatory, called L3, was submitted in 2017. If approved, it would be built and deployed on a four-year mission in the 2020s.

Gravitational waves are 'wiggles' or fluctuations in the fabric of space-time. Collisions of massive objects such as black holes send out these waves. They are also given off by supernova explosions, although those are harder to detect, particularly from Earth. Because of this, astronomers are planning to deploy the LISA L3 observatory well away from Earth. This should make the detection of these faint but fascinating signals from distant collisions much easier. It would consist of a set of detectors set in an array with 2.5 million kilometres of space in the arms between each receiver. With enough data from collisions and other super-energetic events, astronomers will gain new insights into the characteristics of black holes, collisions of close binary stars, and other massive objects.

To the Planets and Beyond with Robotic Probes
The age of solar system reconnaissance has shown us the characteristics of all the planets, comets, asteroids and moons.

There's still a great deal of exploration to be done, from the Sun to the outermost reaches of our neighbourhood in space.

Studying the Sun

Our own Sun is the closest star we have to study, and it continues to challenge solar physicists to explain its behaviour and inner workings. Everything they learn helps them understand other stars, too, which makes the Sun an incredibly important object to study.

Solar Orbiter

ESA's Solar Orbiter (SolO) is set for launch in late 2018, planned to go into an orbit that will take it close to the Sun every five months. The idea is to make very in-depth studies of the heliosphere (the sun's 'atmosphere'), and repeatedly study the same spots each time. A long-term close-up view should be useful through the Sun's activity cycles.

Although astronomers have studied our star for more than a hundred years, many questions remain unanswered. How does the solar dynamo (the process that generates its magnetic field) work? What mechanisms drive the solar wind? What processes combine to heat the corona to incredibly high temperatures? During the Sun's active cycles, prodigious amounts of particle radiation are produced, and solar physicists are working to understand how solar eruptions create these. SolO's instruments will analyse the solar wind and energetic particles, and measure the magnetic field. Studies of the Sun not only help us understand how its activities affect the solar system, but give sharper insight into other stars, too.

Parker Solar Probe

NASA has had its own Sun-orbiting mission on the drawing board since 2008. Originally called Solar Probe, it was renamed in 2017 after Eugene Parker, a solar astrophysicist who proposed that the solar corona could be heated by continual 'nanoflares' (miniature versions of the normal flares the Sun emits) bursting out across the surface of the Sun. It's the first time in history that a mission was named after a still-living person.

The probe, slated for launch in 2018, will loop around the inner solar system for seven years, getting gravity assists at Venus to bring itself into the proper orbit to study the Sun. At its closest, the probe will pass within 6 million kilometres at a speed of 200 kilometres per second. Its scientific instruments will take data that will help solar physicists trace the flow of energy from the solar surface to the corona, determine the structure and dynamics of the solar plasma and magnetic fields that give rise to the solar wind, and study how and why energetic particles are ejected from the Sun. The end result of Parker's mission will give new data on solar activity and provide new information that will help scientists assess the solar wind and predict space weather events that can affect us and our technology.

Earth Sciences and Observations

Perhaps the world's most fascinating observational target is Earth itself! Since the earliest days of the Space Age, orbiting observatories have studied our planet, its atmosphere and its climate. Remote sensing satellites continually return a wealth of data about Earth's geography, changing landforms, hydrology (water cycles), glaciers and geology among other things – including the human effect on Earth's systems. NASA, for example, has continually deployed satellites in its Earth Observations System to answer fundamental questions about how our planet's climate is changing. (There is a list of Earth-observing climate and weather satellites in Appendix F.)

Unfortunately, due to political pressures, NASA and other agencies that study these topics are threatened by those who choose not to understand or accept the climate crisis the planet has entered. Fortunately, future satellites are planned by the ESA, China and others to continue Earth studies.

The interest in our planet's climate is not simply scientific, although the science supplied by current and future satellites will help us understand more about the changes we face. The economic sector, particularly companies that depend on Earth-observing satellites, is also concerned. Private companies such as Finland's Iceye are planning satellite systems to continue their business of watching a changing Earth. Their data will find a home in both government and private labs, serving scientific, agricultural and other interests.

Until relatively recently, one understated aspect of climate change has been its effect on national security. In the US, the military has warned the administration that climate change will quickly become a national security nightmare. Climate change, at its worst, will affect coastal areas, where millions of people live and work. We have already seen what heavy flooding due to warming-influenced rains does to populations throughout the world, including parts of the US during ever-stronger hurricanes. The military considers it a priority to continue efforts to mitigate climate change. This is where data from the world's collection of Earth-sensing satellites can provide the most informed insights.

Near-Earth space is a wide-open arena for other ventures. Orbiting hotels and resorts are certainly one way to raise money from wealthy tourists in the short term, possibly to help fund other scientific endeavours in space. This may seem fanciful but we are likely to see them eventually, once the financial and legal frameworks are established. We've already looked at mining operations on the Moon and asteroids, but some far-future ideas include mining comets for water and towing asteroids back to Earth orbit as bases for colonies or science labs.

An audacious plan from the Chinese Academy of Sciences and Academy of Engineering proposes an orbiting solar collector structure that would generate up to 100 megawatts of power across an array perhaps as large as 6 square kilometres. The power would then be beamed to Earth via a laser or microwave link. It's an ambitious idea that would require hundreds of rocket launches to loft the needed construction materials to space, and take perhaps a decade to complete. The project would almost certainly boost the fortunes of solar power technology companies.

The idea of solar arrays in space is not exclusive to China. The US and Japan are also exploring such stations as a path toward alternative energy use and energy independence. If these projects come to pass, they would certainly be part of the overall solution for global warming and climate change currently being addressed by most nations.

On to Mars

Dreams of trips to Mars are as old as Edgar Rice Burroughs' tales of John Carter and Dejah Thoris. So many science fiction writers have told tales of the Red Planet that it's become a stock adventure trope in cinema. Yet from our dreams come reality, and the promise of a trip to Mars *is* getting closer.

In the next chapter, we'll take a closer look at some of the proposals floating around to take people to the Red Planet. There are still many questions to be answered about the endeavour: who will go? How do they get there? How long will they stay? What will they do there?

The Red Planet remains firmly fixed in our sights. Many missions have been sent there already, and more robotic probes are in the planning stages. The missions that survived the trip to report home have been the equivalent of scouts of pioneer days in the US and the mariners of an earlier age of Earth exploration. They returned reports on conditions in the territory ahead. From the rovers, landers and mappers, humanity has received incredibly tantalising views of a world so much like our own but also very, very different. The more planetary scientists can learn about Mars before the human missions arrive, the better off those new Martians will be as they explore.

It makes sense to send a probe to collect Martian rock and dust samples and return them to Earth. Samples of Martian soil would reveal more about the surface chemistry, and the contributions it makes to the planet's incredibly thin atmosphere. Some soil sampling studies were carried out by the Viking landers, as well as later probes such as Curiosity. However, those tests are limited by the technology that can be sent on a spacecraft. A better geological study can be carried out in labs here on Earth. While some analysis has been done on Martian meteorites that have fallen to Earth, those represent a Mars of the very distant past. 'Modern' Mars rocks should allow a more comprehensive evaluation of conditions in the present and give a much better idea of what resources are available for use once humans get there.

For example, perchlorate is a chemical compound made up of chlorine and oxygen, and is known to exist on Mars. Armed with a better understanding of what those soils contain, future Mars trips might plan to use the perchlorate to help generate breathable air. Perchlorate is also an ingredient in rocket propellants; if enough could be found, it might be used to create fuel for return trips to Earth. It also turns out that there's plenty of subsurface ice and carbon dioxide on Mars, which could also be turned into rocket fuel.

Upcoming missions include NASA's Mars Insight (2018), the Mars 2020 probe, and China's rover, lander and sample return planned for the 2030s. China will use its experience based on a similar mission to the Moon to gather samples and get them back to Earth for analysis. The Mars 2020 rover will drop to the surface using a 'sky crane', as Curiosity did, and then roll around in the Syrtis Major region drilling into rocks to collect samples of minerals and dust. It will 'cache' the samples for a future retrieval mission. The science from these probes should add greatly to the store of knowledge needed for humanity's next steps to Mars.

Jupiter

The giant planet Jupiter has come in for an incredible amount of study, with the Pioneer spacecraft in the 1970s, and the Voyagers, Galileo and other missions. The most recent to visit is the Juno spacecraft, which has given incredibly detailed views of the planet.

It's no surprise that astronomers keep wanting to go back. Jupiter itself has been an enigma, with its massive cloud decks and storms. However, it also has intriguing moons, including icy Europa, which may be hiding a habitable zone beneath its frigid crust. Io, the volcanic moon, continually resurfaces itself with rocky, sulphurous lavas. The radiation environment is deadly to humans, but spacecraft can continue to explore the planet, its moons and its ring system.

ESA is planning a 2022 mission called JUICE – the Jupiter Icy Moons Explorer – to study the four largest natural satellites, Europa, Io, Ganymede and Callisto. It will also carry out detailed observations of the turbulent atmosphere of Jupiter, measure its magnetic field and explore its rings.

In Arthur C. Clarke's novel *2010: Odyssey Two*, the focus of action is on Jupiter's moon Europa. It appears to have life, and the Earth-based explorers who come to the Jupiter system get a chilling warning: 'All these worlds are yours – except Europa. Attempt no landings there.'

Clarke long suspected that this tiny moon could be a safe haven for life, and many planetary scientists think the same thing. For starters, Europa has a deep salty ocean beneath its icy crust, heated from the interior. A mission called Europa Clipper has been planned to make multiple flybys of this tiny moon, and perhaps even fly through plumes of material being ejected from beneath the surface. It would make observations of Europa's outer ice shell, confirm the existence of the deep ocean, characterise the ocean's composition, and find evidence of interaction between the ocean and the surface. The mission would also seek to understand the geology of this tiny world. There is a great deal of interest in figuring out whether human missions could be deployed on Europa in the far distant future. Since this moon is embedded within strong radiation belts surrounding Jupiter, it's unlikely that humans could safely explore there. However, a specially hardened set of instruments on a spacecraft and nanosatellites and smaller probes deployed from the main craft could survive long enough to make a detailed study. Europa Clipper is still in the design phase, so it's not likely to fly until the mid- to late 2020s.

Pluto and Beyond

The continuing quest to understand the worlds of the solar system has already opened up a new region for study: the Kuiper Belt. This is a part of the solar system that, until a few decades ago, was a chilly mystery. Stretching out from the orbit of Neptune, it contains dwarf planets, smaller worlds and countless icy comet nuclei. Pluto is the best-known of the Kuiper Belt objects, and although it's not the largest of the worlds out there, it is the most distant world a spacecraft has visited. New Horizons, which is on its way to explore another object in the Kuiper Belt, flew past Pluto in July 2015. It arrives at its next destination on 1 January 2019.

The outer solar system has a wealth of information locked away on its worlds and in its cometary ices, clues to the origin and evolution of the planets, moons and icy worlds. Unfortunately, no new probes are being built to continue exploring these distant outposts. However, there are plans for a possible Pluto orbiter, or perhaps a probe to study one of its siblings, the dwarf planet Eris. Instead of flying by and grabbing images and data along the way, an orbiter would provide long-term studies of its target. That would require a hefty spacecraft and heavy-lift rocket, such as the Space Launch System, to give it the initial boost of speed it needs to go the distance. After seven years, the probe, perhaps powered by an ion-thrust booster similar to the Dawn spacecraft at Ceres, would arrive at its target and do for Pluto or Eris what the Cassini Mission has done for Saturn over the past decade. That outer solar system mission, which is still very much in the early planning stages, could be approved and launched by the late 2020s or early 2030s. It would arrive at its target seven or eight years later, and begin sending back a treasure trove of scientific data.

Space Exploration Comes Home

In chapter 6, we looked at the 'spinoff' technologies developed for space exploration that also find their way into our homes and businesses. That kind of technology transfer is bound to continue as humans move to space, whether it's an actual bit of space hardware that is being repurposed for use on Earth, or something derived from one.

Robotic mining machines come to mind as one technology that could replace the risky jobs humans now do deep inside Earth. Autonomous trucks, drilling rigs and underground transports are already used in some mines, and as space-age technologies improve, there will be more. Mining is a risky business, and it's likely some jobs that might have gone to underground workers will instead go to those who create and programme robots.

Beyond autonomous vehicles for heavy industry, the self-driving car is an idea whose time has come. These used to be thought of as something 'in the future'. Today they exist, equipped with GPS-enabled navigation technology and artificial intelligence. Once societal attitudes towards the safety of such systems improve, it's very likely carmakers around the world will offer them.

Space Medicine and You

Space exploration has come home to boost medicine in a big way. We've read elsewhere about improved medical sensors and methods for cancer detection influenced by life support and medical systems used in space. What if the venerable *Star Trek*-style tricorder could be invented?

In 2017, a team of researchers and inventors at Final Frontier Medical Devices won a $2.5 million prize from the Qualcomm Tricorder XPrize committee for their DxtER device that uses sensors and computer databases to diagnose a patient's condition. The Dynamical Biomarkers Group took second prize for their DeepQ Kit, which performs similar actions to sense a patient's maladies. In many places, people use small, wearable devices (such as the FitBit) while exercising or simply going about their daily routines. In the future, these systems, based on space technology, will be woven into our clothes and placed around our homes.

Today, doctors practise medicine through 'telepresence', using the internet to reach patients remotely. Already, some physicians use tiny cameras originally developed for space applications to peer into their patients. Surgeons use Microsoft's HoloLens headset to 'see inside' their patient in real time as they operate, while others use robotic devices for precision surgery.

The Far Future

When we look at the far future of space exploration, we can imagine orbiting cities high above Earth, called O'Neill colonies (named for Gerard K. O'Neill, who came up with the idea). They would have homes for hundreds of thousands of people, enjoying Earth-like gravity and weather but inside sealed 'cylinders' occupying orbital space. Or, we can ponder those lunar retirement villages Robert A. Heinlein described in his predictions of the future. The idea makes sense – living in a low-gravity environment might seem a boon for aged but still healthy adults – as long as they have the wherewithal to afford such a distant home. (They would probably be dug deep underground to protect the inhabitants from the relentless fourteen and a half days of solar radiation each month.)

Of course, the futuristic idea of space hotels and resorts keeps cropping up. They're not so far-fetched, but none have been built yet. The test run of a Bigelow expandable module on the ISS shows that at least one piece of the technological puzzle exists to build a hotel. However, the logistics are difficult. Until there's more than one way to get to space, building the hotel is a challenge – as is transporting visitors for orbital vacations.

Space Planes

Access to space is, for the moment, largely limited to rocket-and-capsule technology. The space shuttle programme was supposed to change that, and it did – until 2011. The idea was to provide winged access to space. In the wake of the shuttle fleet's move to museum exhibit status, companies have been looking anew at the concept. Sierra Nevada's Dream Chaser, a smaller version of the shuttle orbiter, is a reusable cargo system that can launch on a rocket and return to a rolling landing on Earth. The company does have a contract to deliver cargo to the ISS for NASA, but the future prize would be a passenger version of the orbiter. ESA and OHB Systems ENG in Europe are studying future use of the Dream Chaser in both human missions and supply sorties to space. There have even been discussions about using the orbiter on a trip to refurbish the Hubble Space Telescope in the 2020s. While it can carry up to seven people, a Hubble refurbishment would require more spacewalks to the telescope than were done in previous servicing missions. The idea behind the suggestion is that if the James Webb Space Telescope has problems, having a 'fix' for Hubble would allow the ageing telescope to function even longer than it already has.

Dream Chaser isn't the only space plane out there, although it has gone through test flights and is ready for missions. India tested its reusable launch vehicle demonstrator in 2016, sending an unscrewed space plane on a short flight as a proof of concept. The British company Reaction Engines is working on a very futuristic plan for a single-stage-to-orbit space plane called Skylon. It is aimed at lofting payloads and cargo to orbit, but eventually it could carry passengers and crew. Future aircraft could use the same engines used in Skylon to reduce the time passengers spend traveling from one country to another by applying Mach 5 speed to hypersonic transport.

Long-term Space Travel
Let's think even further out. We've explored the ideas of colonies on Mars, but what if we could send people to live and work safely on Europa? Or explore Enceladus out at Saturn? Or perhaps even to fulfil Arthur C. Clarke's dream of a civilisation of miners on Titan, funnelling their precious resources back to Earth? This is science fiction for now, but not beyond the capability of humans to achieve. It simply takes time, money and the will to succeed.

What about going to the stars, as astronaut John Young once famously exclaimed after his first shuttle mission? Is that doable?

The far future could bring humans to interstellar space, perhaps to explore those many exoplanets Kepler and other telescopes are discovering. There are plans to send solar sails to nearby stars, traveling at close to the speed of light, to study worlds that may exist there. However, even those missions would take decades or centuries to get to their destinations.

This illustrates an immutable law of nature: the speed of light – the fastest speed we know of – is the limit. We can't go faster. If ships could travel even near that speed, voyages of centuries would only take us to the nearest stars. Trips across the galaxy would take many times longer.

Antimatter rockets have often been suggested as a possible way to accelerate humans on long voyages of discovery. The idea of using antimatter is an interesting one. In theory, a tiny amount of it is needed to set off a controlled energy release strong enough to power a spaceship. When matter and antimatter come into contact, they annihilate in a flash of energy. If that could be captured, then immense amounts of power for rockets would be at our disposal. However, the problems with antimatter are twofold. First, there isn't a lot of it around; and second, we don't have the technology to make it or contain it.

There *were* equal amounts of matter and antimatter at the beginning of the universe, but since then matter has come to dominate the known universe. So, the search is on for antimatter. Scientists at the Large Hadron Collider in Switzerland seek to trap antimatter particles for study, and others are looking at technologies to search out and trap the stuff in space. Even if they find a source, they'll have to contain it and figure out how to build engines powered by antimatter/matter annihilations.

There are other possibilities for fuels. Hydrogen is one. We are familiar with it in its gas or liquid state, but it can and does exist as a metallic substance. It can be made into fuel for rockets by heating the solid to make a gas. That process generates power. The big problem is, where do we find massive quantities of it? It doesn't exist on Earth in a natural state. The planet Jupiter is 90 per cent hydrogen, and much of it is compressed deep inside the planet in a liquid metallic state. However, that's not accessible to us. However, scientists have found a way to make liquid metallic hydrogen in the lab. If it turns out to be feasible for large-scale production, that would supply a whole new way to power our rockets to space.

NASA has been using ion propulsion (a method that ionises, or energises, propellant to provide thrust) on such missions as the Dawn probe to Ceres; it has also been used on a number of other missions. Another way to send probes, particularly around the inner solar system, is through the use of solar electric propulsion. A rocket is launched to space with a propellant that can be heated by solar energy to produce thrust.

Finally, there's warp drive – or, as it's often called, faster-than-light drive. The most promising ideas for this technology come from a concept called the Alcubierre Warp Drive, which is based on a calculation for how to stretch the fabric of space-time. This would produce a wave that would somehow contract space in front of a spacecraft at the same time expanding space behind it. The spacecraft would, essentially, be riding a wave of space-time. The Alcubierre drive would form a bubble in space-time and the ship would be inside. At the present time, there aren't any technologies able to create this wave and allow a ship to enter and leave it at the beginning and end of the trip. There have been experiments related to the concept at NASA's Jet Propulsion Laboratory, but the results so far have not led to the invention of even a test vehicle.

That leaves long-term trips as a way to travel among the stars (or even to the outer reaches of our solar system). Generation ships would leave Earth headed for a destination such as Proxima Centauri (the nearest star to us, at a distance of 4.1 light-years). Along the way, multiple generations of people are born, live and die as the ships travel through space at some fraction of the speed of light. Eventually, the distant descendants of the first men and women to ride those ships will finish the trip.

Another possibility is for the crews to travel in hibernation. Centuries later, they wake up at their destination and send a message back to confirm arrival. Hibernation and generation ships are tools of civilisations ready to go out and explore the distant reaches of the galaxy. For humanity, they're in the far future, to be used after we've explored our own solar system more thoroughly, and made settlements in near-Earth space, the Moon and Mars.

Stepping into the Future

The promise of space exploration has beckoned to humanity for more than a century, through the future histories of science fiction or the realities of science fact. The early launches, the race to the Moon, the reconnaissance missions to the planets – all were just the first steps. In a sense, humanity has spent this past century working out the details of flight, on Earth and then off our planet. Valuable lessons have been learned and, to be sure, painful mistakes have been made.

The Space Age began as a product of war, whether we like to think of it that way or not. Goddard's first rockets may not have received much respect, but it didn't take long for generals and politicians to see the benefits of having superior technology during times of conflict. Of course, rockets today are still primed to deliver warheads, but they also enable satellite communications, lunar probes and orbiting observatories, and will eventually enable trips to the planets.

How the Space Age will evolve is an open question, and it's up to the interested countries of the world to answer. One thing we know is that it is not going to end any time soon. Most of the world has become quite used to the 'amenities' of the Space Age – the smartphones, improved medical devices, ease of flight around the world, and so much more. As a species we are more connected than ever before, with nearly instantaneous communication, sharing and learning thanks to space-age technologies. Where we go with it from here is entirely up to us.

10

Mars Quest: Humanity's Future on the Red Planet

Mars has been flown by, orbited, smacked into, radar examined, and rocketed onto, as well as bounced upon, rolled over, shoveled, drilled into, baked and even blasted. Still to come: Mars being stepped on.

Buzz Aldrin

We need to be laser-focused on becoming a multi-planet civilisation. That's the next step.

Elon Musk, SpaceX

The story of space exploration is an ongoing adventure. It may have begun with the leftover materials of war being refitted for scientific endeavour, but today it is an important piece of humanity's future. Those who have grown up with the Space Age can scarcely imagine a world without launches and orbiting space stations and future trips to the Moon and beyond. Now, new generations of explorers are planning missions to near-Earth space and beyond.

The most talked-about mission of the future is that of a human trip to Mars. NASA, ESA, Roscosmos, CNAS, ISRO and private companies such as SpaceX have all envisioned sending people to Mars. Of course, science fiction readers have explored Mars in their imaginations since before the turn of the twentieth century.

Before people go to Mars, however, they'll likely visit the Moon again. Compared to Mars, that's an 'easy' trip. It's only a few days away, and a handful of people have already been there. We know how to do it. As mentioned in chapter 9, the Chinese and Europeans are looking toward a lunar village, and NASA has plans to return there. It makes a lot of sense to go back, to learn how to survive on the Moon as a precursor to Mars trips. Of course the two worlds aren't much alike, but the Moon lets people test new technologies, build infrastructure and learn to work in protective environments similar to potential Mars bases.

The best part about returning to the Moon is that, if anything goes wrong, possible rescue is only a few days away at best. Still, there is a nagging sense of 'been there and done that'. That has contributed to pressure from some quarters to skip the Moon and go straight to Mars, if not somewhere beyond.

There aren't too many places in the solar system where humans can live and work as they do here on Earth. While a trip to Europa makes for a great science fiction story, the reality is far less benign. That little moon, embedded in Jupiter's radiation belts, would subject a visitor to a deadly dose of energetic particles that are difficult even for a well-shielded spacecraft to withstand. Venus, which is a terrestrial planet, is nowhere near safe enough to live on, although there have been some far-future ideas about building orbiting stations there. Closer to home, the idea of O'Neill colonies orbiting Earth have some cachet, but their implementation is pretty far in the future. So, we're left with Mars as a future exploration opportunity.

The Logistics of Mars

A trip to Mars is not a quick jaunt. It most certainly isn't an Apollo-style 'plant the flag, grab rocks, come home' kind of expedition. Sending people there is going to be an expensive, long and technologically challenging project. With respect to those who say we have all the technology we need 'on the shelf' to do it, there are still ships to be built, tested, rebuilt and retested. We haven't flown beyond the Moon, and the risks and problems of long-term space flight beyond the Moon are considerably more complex.

However, problems can be solved. It won't be long before the first Martians are on their way. Once there, it makes more sense for them to stay on Mars to carry out scientific studies, explore and establish places to live and work than to rush home again and wait for the next group of travellers to do the same thing.

In a 1984 conference called 'The Case for Mars', mission planners and strategists set forth an ambitious plan of action. Here, in brief, is what it contained. First, there would be precursor missions sent to the planet. They would be rovers, mappers and landers. These missions would gather advance information about the weather and the long-term climate, the geology, resources that could be used for propellants, and the availability of water.

While those 'advance scout' missions were underway, people on Earth would be using Mars-analogue landscapes to train the future 'Mars-nauts'. NASA, ESA, Roscosmos and others involved in the Mars missions would get busy building an orbiting space station where the trans-Mars ships would be built. The materials for those spacecraft would come from lunar mines operated by multinational interests. Once all the pieces were in place, the crew members (from around the world) would board their craft and take off on the months-long journey to a new world. Their goal: science exploration and the eventual creation of labs, habitats and cities. At the time, all this was planned to take place perhaps in the late 1990s, or the early decades of the twenty-first century. It all looked very straightforward back then.

Amazingly enough, given the vicissitudes and vagaries of funding and politics, much of that plan of action has occurred. Perhaps it hasn't happened as quickly as originally intended, but there has been progress. Decades later, the Mars rovers and mappers have been sent. Most have been wildly successful, and the scientific gains are tremendous. Thanks to them we know a lot about the planet, but there's much more to learn. As for the rest of the early 'Case for Mars' scenario, we've only had a few space stations, and they haven't been ship construction sites. Nor are there any lunar mines – yet.

What's on the Drawing Board

There is progress towards eventual human missions to the Red Planet. NASA has proposed the 'Journey to Mars', a series of missions that continue the robotic exploration for at least another decade. A series of ever more complex missions will allow astronauts to train in deep space, work on asteroid missions and practise many of the skills needed for a longer mission to the Red Planet.

China's space agency is focused on Mars as well. Their first proposed mission to the Red Planet is a lander, rover and orbiter combo that would join the search for any possible life, study the surface and environment, and analyse the composition of the rocks and soil. Human missions would follow, but on a timetable yet to be released.

The Russians announced in 2016 they could send a nuclear-powered spacecraft to Mars in forty-five days. That would nearly halve the travel time of most other missions, but it's not clear where they are in the development of that ship.

In 2016, SpaceX chairman Elon Musk said he wanted to see a million people living on Mars by the end of the twenty-first century. Of all the people on planet Earth, he and his company might be the best-positioned to send colonists on eighty-day journeys to the Red Planet, to settle and make homes on a hostile world. A trip to Mars is, after all, what set Musk on the path to creating SpaceX and its profitable launch empire. The key for his dream to come true is *returnability*. There have to be multiple trips carrying hundreds and hundreds of people. Once they get there, those people have to adapt to the conditions, build their homes and procreate – the latter perhaps being tricky in such a new environment. Producing a viable new generation on Mars represents a huge challenge.

SpaceX and NASA are not the only ones to look with longing on a distant world and dream of future cities there. India has already sent its first Mars probe and China is building one. Whether Russia, India and China will then send people is an open question. One thing that's not in doubt, however, is that the colonisation of Mars is on a lot of minds.

While NASA and SpaceX are planning missions, others have offered up some 'fast-track' approaches. The Mars One group, led by entrepreneur Bas Landorp, for example, would send supplies, rovers and other equipment to Mars ahead of any human missions. While those 'supply trains' are in progress, Mars One mission teams will be training for their eventual trip. Then, when everything is in place, the human crews head to Mars. Once there, the first crew sets up while the second crew makes its way two years later. The Mars One group has come in for a fair amount of criticism for its business model, its preliminary crew selection policies and its all-or-nothing approach, where it expects people to go to Mars but not necessarily come back. The idea is that Mars One will continue to send missions, with each crew welcoming the next as they build their colony. The company is still refining its business plan. The mission scenario calls for the first human missions to arrive in 2031.

Engineer Robert Zubrin's 'Mars Direct' mission scenario takes a similar approach to the Mars One idea, proposing a fast-track mission using off-the-shelf technologies. He wants to get people to Mars as soon as possible in a series of missions. The mission would send an Earth-return vehicle (ERV) to Mars, where it would land and deploy a rover with equipment and supplies to generate rocket fuel for the ERV. While that happens, a second ERV is sent along with astronauts in a habitat. They arrive at Mars, do a year and a half of exploration, and then return to Earth on the fuelled ERV. In the meantime, a second ERV and habitat are on the way to Mars to continue the mission. Eventually, there are enough habitats and rockets and other equipment to service a growing human settlement (or more than one) on the Red Planet.

Challenging a New Generation

No matter what the first human missions to Mars are, they will very likely be international collaborations, if for no other reason than to spread out the incredible cost of sending them. The first generation of Mars explorers may already be walking among us – our children, nieces, nephews, students, friends. If SpaceX gets its missions going, some Mars colonists will pay a lot of money to be among the first to set foot there. No matter how they get to Mars, from the moment they arrive human Martians will become a society radically shaped by the environment of Mars. What will they see? What will their children be like? Their civilisation?

While space agencies and companies develop the technology to go to Mars, others have to plan for the challenges Marsnauts will face, in particular their health and mental stability. The long-term space journeys they undertake will affect their bodies in many ways, from musculoskeletal changes to altered fertility, vision, hearing and balance. Recent studies show Marsnauts could face increased risks of cancer, and the radiation would affect any pregnancies.

Long-term inhabitants in the microgravity of the ISS have shown effects that could deal a serious blow to months-long voyages to

Mars. We already know from the experiences of ISS astronauts that extended stays in low-gravity environments adversely affect their bodies. Even if a spacecraft with a 1G habitat could be built, there remains the question of the psychological adaptations people will need to make to the enclosed environment. However, the ISS experience, coupled with studies of military personnel aboard submarines, seems to suggest the risk of mental and emotional problems is better understood than the physical issues.

All those issues face travellers on the journey. Once they get to Mars, they will confront new conditions that are sure to change them in many ways. Like all missions, a trip to Mars is fraught with dangers and unknowns – while at the same time presenting opportunities for scientific and technological advances.

The Challenges of Living on Mars

As we explored elsewhere in this book, the first explorers to the Red Planet will have some serious challenges ahead of them. Their bodies – adapted through millions of years of life's evolution on this planet – will have to adjust to a very different environment. Conditions on the Red Planet are far from benign. Living there will require constant vigilance and shelter from intense ultraviolet radiation. No one will ever be able to walk outside without a space suit. They will have to grow their own food (fans of the popular movie *The Martian* know that potatoes grow well there), and find their own water. Their diet will be radically different from what we eat on Earth. For the new Martians, life will be very much like living on a submarine, but with windows to a new world.

People who go to Mars will stay there for months, or perhaps the rest of their lives. That means population growth. For new generations of Mars inhabitants, the low-gravity environment they grow up in will *not* prepare them for a trip to the Earth their parents knew and loved. In only two generations, the human race will have diverged – perhaps to become *homo sapiens* and *homo caelestis* (people shaped by the environment of space and lower gravity than Earth).

What will Mars societies be like? Will Earth's politics be transplanted? Its religions? Its social mores? Or will a new type of society arise, changed and shaped by the very harsh environment it occupies? The answers to those questions have long been staples of science fiction, but in a few decades they'll have real answers from real people living radically different lives to those we know here on Earth.

Science fiction stories have shaped our perception about what it will be like to live on Mars, long before Mars missions provided actual glimpses of the planet. In *Red Planet*, Heinlein told a tale of Mars students encountering indigenous Martian 'elders'. The twin boys in *The Rolling Stones* get into no end of trouble during a visit to Mars, where they run into a native Martian. Heinlein's *Podkayne of Mars* tells a story of political intrigue set against a backdrop of solar system colonies on the Moon and at Venus. Of course, many readers are familiar with his *Stranger in a Strange Land*, the tale of a boy born on Mars who comes back to Earth as the sole heir to the entire planet.

Perhaps some of the most evocative stories of humans on Mars were told by Ray Bradbury in *The Martian Chronicles*. They begin with the first colonists to Mars and the reactions of the indigenous Martians, and end with human 'Martians' exploring the only home they have left after Earthlings destroy themselves and the home planet in a bitter war.

Author Kim Stanley Robinson's acclaimed *Mars Trilogy* tells a more gritty and realistic story of humanity exploring the planet. Eventually people in his story transplanted many of the same political and religious issues to a new locale. That's probably to be expected – the first humans to Mars (or anywhere) can't just leave their cultural expectations behind. But in the multinational crews of the future, people will need to figure out a way to coexist in environments so alien that mutual dependency will outweigh religious or political differences.

Perhaps the reality of future Mars missions lies somewhere between the extremes, and people's visits will be more like Andy Weir's *The Martian*, with explorers who just settle down to get the job done despite the obstacles and challenges. (However, most settlers would not want to experience his five years of isolation eating potatoes.)

The 'can-do' spirit is something humanity has in abundance, and no matter who gets to Mars or the Moon or the moons of Jupiter first, the Space Age will continue well into the future. The history of space exploration in this book outlines our marvellous achievements and also our failures. The future in space provides both exciting opportunities and the prospect of a lot of work for a lot of people. That's the glory as well as the pain of becoming a spacefaring civilisation.

Where Do We Go from Here?

Some years ago, the full-dome production company Loch Ness Productions produced a planetarium show called *MarsQuest*. It played in domes around the world, keeping the spirit of Mars exploration alive for audiences everywhere. It imagined a future Mars teeming with people, descendants of the first Mars explorers, told from the viewpoint of a person who had spent a lifetime on Mars. Looking back at the history of the planet's exploration, he stands in front of a window overlooking the Marsopolis so many of his fellow Mars inhabitants call their homes, reminiscing aloud about the long, strange journey that brought humans to the Red Planet. His closing thoughts in the show echo the story of space exploration, of humanity's desire to go beyond our planet:

> *Mars! A philosopher once proclaimed our quest to explore your surface as the greatest achievement of the human race.*
>
> *You are so much like Earth — and yet so different. Existence on your surface has never been easy.*
>
> *But, always, we dreamed of walking your dusty plains.*
>
> *Our first steps were small ones, taken by robotic probes that stirred your surface and sniffed your atmosphere.*
>
> *The efforts of our early explorations only whetted our appetites for more. Before we knew it, our first explorers were on the way.*
>
> *The hopes of humanity followed those explorers across the gulfs of interplanetary space. And when they took their first steps onto the Martian surface, it was the end of an old era, and the beginning of a new.*
>
> *These pioneers tested themselves against your desolate wilderness. They charted the broken terrains of Noctis Labyrinthus...*
>
> *and Cydonia Mensae...*
>
> *They crossed Syrtis Major...*
>
> *and mapped Nirgal Valles...*
>
> *They hiked Candor Chasma...*
>
> *and soared over Utopia Planitia...*

At last, they scaled the heights of Olympus Mons, and cast their gaze onto the new frontier.

Mars! In the infancy of humanity, we feared you as a God of War.

Then, we strove to learn your mysteries, and conquer your beautiful desolation.

Finally, the restless wanderers among us made you our second home.

Earth is now a faded memory – a pale blue dot shining in our night sky.

But our Martian cities stand as a tribute to all the dreamers of Earth.

Now, we are the Martians.

That's where space exploration is taking us – beyond our home to a new place where humanity can take the next step to becoming that spacefaring society we've been preparing to become for more than a hundred years.

Glossary

While most terms are defined as they are used in this book, this glossary provides further explanations.

Aeroshell A heat-shielded rigid outer covering for a spacecraft. It protects the vehicle during entry into a planetary atmosphere, shielding it from high temperatures, pressures, and debris (dust, etc.).

Antimatter (*see also* Matter) In regular matter, atoms are made of neutrons, electrons, and protons. Antimatter is made up of the anti-particles that are partners to those in matter, but with opposite electric charge and other characteristics. When matter and antimatter meet, they annihilate each other.

Arcsecond In angular measure, arcseconds, arcminutes and degrees are used to locate an object's location in space. Each arcminute is 1/60 of a degree; each arcsecond is 1/60 of a minute. Used in astronomy and space navigation.

Astronomy The science that focuses on the study of objects beyond Earth, including solar system objects, stars, galaxies and the interstellar medium. There are several subdisciplines of astronomy, divided by regimes in the electromagnetic spectrum or specialisation:

Gamma-ray astronomy: The observation of gamma rays at photon energies above 100 keV (kilo-electron-volts); this is the most energetic form of electromagnetic radiation to be detected from naturally occurring objects and events in the universe. Gamma-ray astronomy is carried out largely from space because the incoming radiation is absorbed by Earth's atmosphere.

Infrared astronomy: The branch of astronomy that studies astronomical objects that radiate, reflect or emit infrared light between 0.7 and 300 micrometres. Infrared is sometimes referred to as heat energy. Much of the infrared light is absorbed by Earth's atmosphere, so high-altitude and space-based observatories are required.

Microwave astronomy: The study of naturally occurring objects and events that reflect or emit microwaves (high-energy radio waves). It encompasses the range between 1 and 100 GHz (gigaHertz). Microwave astronomy is best done from space-based observatories.

Optical (visible) astronomy: The scientific study of objects that radiate so-called 'visible' or 'optical' light, that is, in the visible range of the spectrum that our eyes also detect. The range is roughly 400 to 700 nanometres.

Radio astronomy: The scientific study of objects that give off radio waves, generally between 13.36 and 1427 MHz (megaHertz), sandwiched in between broadcast and other bands.

Ultraviolet astronomy: The scientific study of objects that radiate electromagnetic radiation at ultraviolet wavelengths between approximately 10 and 320 nanometres. Ultraviolet is largely absorbed by Earth's atmosphere and requires space-based observatories.

X-ray astronomy: The scientific study of objects in the universe that emit x-rays. Since they are absorbed by Earth's atmosphere, observations must be done by very high-altitude balloons or space-based observatories.

Atmosphere   A layer of gases surrounding a planet, moon or other object, held in place by the gravitational influence of that body.

Binary   In astronomy, a binary consists of two objects in a common orbit; the objects can be asteroids, black holes, neutron stars and stars.

Black Hole   An object with a gravitational field so strong that neither matter nor light can escape it. The central body of the black hole is a singularity, surrounded by an accretion disk of matter feeding the black hole.

Caltech   The California Institute of Technology, Pasadena, California, a private university that manages NASA's Jet Propulsion Laboratory for the agency.

Capsule   A spacecraft used to carry humans and/or supplies to space. It is usually mated to a rocket that boosts it to orbit.

Celestial body   Any naturally occurring object or body beyond Earth's atmosphere.

Centrifuge for high-G training   A spinning device used in astronaut training to simulate higher-than-normal gravitational forces (G-forces) on the human body.

Comet   A small chunk of ice, rock and dust originating from the outer reaches of the solar system. When it passes near the Sun, it warms and the ices sublimate and flow away from the comet nucleus. This creates a dust tail and frequently a plasma tail.

Communications satellite   An orbital satellite that relays communications signals from one point to another on Earth.

Doppler shift   *See* Redshift.

Drag   (For flight) The resistance provided by the atmosphere as a plane or rocket moves through it. Orbiting satellites also experience 'atmospheric drag' as they encounter Earth's blanket of air. Drag acts to slow an object.

Electromagnetic spectrum   The range of all types of electromagnetic radiation, from microwave, through radio, infrared, visible, ultraviolet, x-ray and gamma rays.

Galaxy cluster   A collection of galaxies bound together by gravity.

Gravity   The force of attraction between two or more masses in the cosmos, from particles to large-scale structures (such as galaxy superclusters), and including dark matter.

Hertz (Hz)   A unit of frequency. One hertz is one cycle per second. Kilohertz (kHz) is a thousand cycles per second; megaHertz (MHz) is a million cycles per second. Radio astronomy uses these terms.

Hydrocarbons   Compounds of the elements hydrogen and carbon; they are found in petroleum products such as methane (CH_4) and ethane (C_2H_6), which are also found in the atmosphere of Titan at Saturn and the distant planet Pluto.

In situ   A Latin term meaning 'in the natural position or place'; in planetary science, used to mean studying something 'on the planet' in its original environment.

Ion propulsion drive/ion thruster   A form of electric propulsion that uses electricity to accelerate ions. Designs have existed and been tested since the 1970s. Among the spacecraft using this technology are NASA's Dawn mission, Japan's Hayabusa and ESA's Smart 1. Future missions including the technology are BepiColumbo, and possible trips to Mars.

Kelvin   Denoted K, a unit of temperature. The Kelvin scale (named for Lord Kelvin, who devised the scale). 0 kelvin is equal to -273 C or -459 F. Proper usage is to state the number followed by the word 'kelvin' in lowercase.

Lander   A spacecraft designed to land on the surface of another body (planet, moon, comet, asteroid).

Light-year   The distance light travels in a year at a speed of 299,792,458 metres per second; the total distance is 10 trillion kilometres.

Lunokhod   A robotic vehicle deployed by the Soviet Union to explore the surface of the Moon.

Mach number   Named after Austrian scientist Ernst Mach; a number used in aeronautics to express the ratio of the speed of an aircraft (or spacecraft leaving Earth) to the speed of sound.

Mass   A body of matter, or one of the physical characteristics of a body. The basic unit of mass is the kilogram (kg). An object's mass is also the measure of its resistance to acceleration (that is, a change in its motion). Mass is not the same as weight. An object from Earth would have the same mass on the Moon, in orbit, on Mars, etc., but its weight would be different depending on where it is and the gravitational field it occupies.

Matter   Any substance that has mass and takes up space; matter is made of protons, neutrons and electrons.

Microgravity   When people or objects in space are in a weightless state. In low Earth orbit, for example, Earth's gravitational influence is very small, producing what feels like weightlessness.

Payload   The material delivered to space by a rocket; can also be a weapon delivered by a missile.

Propellant   The fuel used in rocket motors and engines to provide thrust for motion (or to slow the system down). Propellants can be liquid or solid. The liquid forms can be petroleum-based, hypergolic or cryogenic. Kerosene is often used, as is liquid hydrogen or liquid oxygen. Solid-propellant fuels are the earliest types used and date back to the early Chinese fireworks. Once ignited, a solid must burn all the way until it is exhausted; it cannot be stopped and started.

Propulsion system   (In space travel) The entire launch/space vehicle, including the rocket engines, fluid lines, tanks and other equipment needed to provide propulsive force.

Redshift   The apparent increase in the wavelength of light coming from a source as it moves away (or recedes). This is also known as a Doppler shift.

Regolith   A layer of loose material covering solid rock on a planetary or asteroid surface. It may include dust, sand and broken rock.

Rocket   A self-contained projectile propelled by a rocket engine and fuel. Today's space programs use chemical rockets, built in two or more stages. They can be propelled by solid or liquid fuels. Examples of space-based launch rockets are the SpaceX Falcon 9, the French Ariane 5, the Russian Soyuz and the Chinese Long March series.

Rocket engine   A type of jet engine in a rocket that uses propellant to create propulsion.

Satellite   In space flight, an artificial object such as a communications satellite orbiting Earth (or another body); in astronomy, a naturally occurring object orbiting a primary, such as Earth's Moon, or the moons of Mars.

Soyuz   A Russian launch vehicle, the most used in the world. The Soyuz spacecraft that is used to carry humans and cargo to space.

Space age   A period in human history that began as early as the late 1940s and continues to the present day. It encompasses the world's history in space exploration.

Space-based astronomy   Astronomical observations and studies made using orbiting observatories and planetary probes.

Spacecraft   A general term used to refer to vehicles that travel in space; these can carry crew or be robotic.

Space Race   The competition between the United States and Soviet Union to reach the Moon with human crews. It began when the Soviets launched Sputnik 1 to orbit on 4 October 1957 and ended with the American Apollo 11 landing three men on the Moon on 16 July 1969.

Space radiation   Ionising radiation that exists in the form of high-energy charged particles. It comes from three sources: solar particles, galactic cosmic radiation (cosmic rays, etc.), and radiation trapped in Earth's magnetic field.

Space spin-offs   Technologies developed for use in space exploration that are transferred to use on Earth in a variety of uses, e.g. telescope technology transferred for use in medical settings.

Specific impulse   The total impulse delivered per unit of propellant in a rocket; also a measure of the efficiency of a rocket or a jet engine.

Thrust   A reaction force that works when a system (such as a rocket) expels or accelerates mass in one direction; this causes an equal and opposite force.

Voskhod   A three-person spacecraft built by the Russians to carry humans to space.

Vostok   The first spacecraft built by the Russians to carry humans to space.

Countries with Orbital Launch Capability and Their Current Rocket Families

(Correct as of mid-2017.)

Australia
SSRP (sounding rockets)
AUSROC Nano
AUSROC 2.5 (in development

China
(Source: http://www.cnsa.gov.cn/)
Long March 2F
Kuaizhou
Long March
Long March 2C
Long March 2D
Long March 2F/G
Long March 3
Long March 3A
Long March 3B/E
Long March 3C
Long March 4
Long March 4B
Long March 4C
Long March 5
Long March 5B (under development)
Long March 6
Long March 7
Long March 11

France/Europe
(Source: http://www.arianespace.com)
Ariane 5
Ariane 6 (in development)
Vega (developed with Italian Space Agency)

India
(Source: http://isro.gov.in/launchers)
Geosynchronous Satellite Launch Vehicle (GSLV and GSLV Mark III)
Polar Satellite Launch Vehicle (PSLV)
Various sounding rockets
Reusable Launch Vehicle (RLV)

Israel
(Source: Burleson, D., Space Programs Outside the United States*)*
Shavit
Sounding rockets

Iran
Safir-2
Simorgh SLV

Japan
(Source: global.jaxa.jp)
Epsilon
H-11, H-11A, H-11b
H3 (in development)
Various sounding rockets

New Zealand
Electron

North Korea
(Source: http://space.skyrocket.de/doc_lau/unha.htm)
Unha family

Russia
Angara family
Proton-M
Soyuz family

South Korea
Naro family (KSLV)

United States
Antares
Atlas V
Delta II, Delta IV, and Delta IV Heavy
Falcon 9 family
Minotaur and Minotaur-C (Taurus)
New Glenn
New Shepard
Pegasus
Space Launch System (in development)

Human Missions Flown

(Correct as of mid-2017.)

China
First crewed mission: 2003
Shenzhou programme: 2012
Tiangong programme: 2013-present

NASA
Mercury programme: 1959–63, 6 mission
Gemini programme: 1963–66, 10 missions
Apollo programme: 1961–72, 11 mission
Skylab: 1973–74, 3 missions
Apollo-Soyuz: 1975, 1 mission
Shuttle programme: 1981–2011, 135 missions
Shuttle–Mir programme: 1995–98, 9 missions
International Space Station: ongoing

Soviet Union/Russia
Vostok programme:1961–63, 6 missions
Voskhod programme: 1964–65, 2 missions
Soyuz programme: 1973–91, 66 missions
Soyuz programme: ongoing, up to 133 missions, more planned

APPENDIX C

Agencies and Institutes

ASC/LPI: Astro Space Center of PN Lebedev Physics Institute (Russia)
ASI: Agenzia Spaziale Italiana (Italy)
Astrium: aerospace manufacturer/now part of Airbus Group
BMDO: Ballistic Missile Defense Organization (US Dept, of Defense)
CNES: Centre National d'Études Spatiales (France)
CNRS: Centre Nationnal de la Recherche Scientifique (France)
CSA: Canadian Space Agency
DOST: Department of Science and Technology (Phillippines)
DLR: German Aerospace Center (Deutsches Zentrum für Luft und Raumfahrt e. V)
ESA: European Space Agency
EUMETSAT: European Organisation for the Exploitation of Meteorological Satellites
GISTDAL Geo-Informatics and Space Technology Development Agency (Thailand)
IKI: Russian Space Research Institute
INPE: Instituto Nacional de Pesquisas Espacials (Brazil)
INTA: Instituto Nacional de Técnica Aeroespacial (Spain)
ISA: Italian Space Agency
ISAS: Institute of Space and Astronomy, Japan (now part of JAXA)
ISN: Istituto Nazionale di Fisica Nucleare (Italy)
ISRO: Indian Space Research Organisation
JMA: Japan Meteorological Agency
JAXA: Japan Aerospace Exploration Agency
KARI: Korea Aerospace Research Institute
LANL: Los Alamos National Laboratory
LAPAN: Lembaga Penerbangan dan Antariksa Nasional (Indonesia)
MBRSC: Mohammed Bin Rashid Space Centre (Dubai, UAE)
NASA: National Aeronautics and Space Administration
NASDA: National Space Development Agency of Japan

Appendix C: Agencies and Institutes

NOAA: US National Atmospheric and Oceanic Administration
PPARC: Particle Physics and Astronomy Research Council (now part of Science and Technology Facilities Council, UK)
RSRI: Russian Space Research Institute
Roscosmos: Russian state space agency
SERC: Science Engineering Research Council (UK; merged into Science and Technology Facilities Council)
SNSB: Swedish National Space Board
SRON: Netherlands Institute for Space Research
SSC: Swedish Space Corporation
TÜBİTAK: Scientific and Technological Research Council of Turkey
USGS: United States Geological Survey
USN: United States Navy

Timeline of Planetary Explorations

(Sources: NASA, ESA, Roscosmos, ISRO, CNSA)
Dates reflect day and year of launch
Unless otherwise indicated by 'crewed', missions are robotic probes.
Soviet probes designated 'Kosmos' reflect a change in mission status to Earth orbit
due to launch or other difficulties.
** Indicates mission still operational (as of mid-2017)

1957
4 October: Sputnik 1, successful Earth-orbiting mission (USSR)
3 November: Sputnik 2, successful Earth-orbiting mission (USSR)
6 December: Vanguard TV3, intended Earth-orbiting mission, failed on launch (NASA)

1958
1 February: Explorer 1, successful Earth orbiting mission (NASA)
17 March: Vanguard 1, successful Earth orbiting mission (NASA)
17 August: Pioneer 0, intended lunar orbit, failed on launch (NASA)
23 September: Luna A, intended lunar mission, failed on launch (USSR)
11 October: Pioneer 1, intended lunar orbit, failed on launch (NASA)
12 October: Luna B, possible intended lunar impact, failed on launch (USSR)
8 November: Pioneer 2, intended lunar orbit, failed on launch (NASA)
4 December: Luna C, possible lunar impact, failed on launch (USSR)
6 December: Pioneer 3, intended lunar flyby, failed on launch (NASA)

1959
2 January: Luna 1, possible lunar flyby or impact, missed Moon (USSR)
3 March: Pioneer 4, successful lunar flyby (NASA)
6 June: Luna A, possible intended lunar impact, failed on launch (USSR)
12 September: Luna 2, successful lunar impact (USSR)
4 October: Luna 3, successful lunar flyby (USSR)
26 November: Pioneer P-3, intended lunar orbiter, failed on launch (NASA)

1960

15 April: Luna 1960A, intended lunar flyby, failed on launch (USSR)

18 April: Luna 1960B, intended lunar flyby, failed on launch (USSR)

25 September: Pioneer P-30, intended lunar orbit, failed on launch (NASA)

10 October: Marsnik 1 (Mars 1960A), intended Mars flyby, failed at launch (USSR)

14 October: Marsnik 2 (Mars 1960B), intended Mars flyby, failed at launch (USSR)

15 December: Pioneer P-31, intended lunar orbiter, failed on launch (NASA)

1961

4 February: Sputnik 7, intended Venus mission, never left Earth orbit, decayed after 22 days

12 February: Venera 1, Venus flyby, contact lost (USSR)

23 August: Ranger 1, intended lunar test flight, partially successful (NASA)

18 November: Ranger 2, intended lunar test flight, failed second-stage burn, re-entered atmosphere (NASA)

1962

26 January: Ranger 3, intended lunar impact, missed Moon, went into heliocentric orbit (NASA)

23 April: Ranger 4, lunar impact mission, tools failed (NASA)

22 July: Mariner 1, intended Venus flyby mission, failed on launch (NASA)

25 August: Sputnik 19, intended Venus lander, failed to leave Earth orbit, re-entry on 28 August (USSR)

27 August: Mariner 2, successful Venus flyby (NASA)

1 September: Sputnik 20, intended Venus flyby, stranded in LEO, re-entered atmosphere (USSR)

12 September: Sputnik 21 (Venera 2MV02 No. 1), intended Venus flyby, failed to leave Earth orbit (USSR)

18 October: Ranger 5, intended lunar impact, failed to reach target (NASA)

24 October: Sputnik 22, intended Mars flyby, broke up in Earth orbit (USSR)

1 November: Mars 1 (Sputnik 23, Mars 2MV-4), Mars flyby, partially successful (USSR)

4 November: Sputnik 24, intended Mars lander, failed to leave Earth orbit, orbit decayed (USSR)

1963

4 January: Sputnik 25, intended lunar lander, failed to leave Earth orbit, orbit decayed (USSR)

2 February: Luna 1963B, intended lunar lander, failed on launch (USSR)

2 April: Luna 4, intended lunar lander, missed Moon, went into heliocentric orbit (USSR)

11 November: Kosmos 21, possible Venus flyby, stranded in Earth orbit (USSR)

1964

30 January: Ranger 6, lunar impact, cameras failed (NASA)

19 February: Venera 1964A, intended Venus flyby, failed on launch (USSR)

1 March: Venera 1964B, intended Venus flyby, failed on launch (USSR)

21 March: Luna 1964A, intended Lunar Lander, failed on launch (USSR)

27 March: Kosmos 27, intended Venus flyby, failed and burned on re-entry (USSR)

2 April: Zond 1, Venus flyby mission, lost contact with spacecraft (USSR)

20 April: Luna 1964B, intended Lunar lander, failed on launch (USSR)

4 June: Zond 1964A, intended Lunar lander, failed on launch (USSR)

28 July: Ranger 7, successful lunar impact (NASA)

5 November: Mariner 3, intended Mars flyby, lost power 8 hours after launch (NASA)

28 November: Mariner 4, successful Mars flyby, first images of another planet (NASA)

30 November: Zond 2, Mars flyby, lost contact with spacecraft in May 1965 (USSR)

1965

17 February: Ranger 8, successful lunar impact (NASA)

12 March: Kosmos 60, intended lunar lander; failed to leave Earth orbit (USSR)

21 March: Ranger 9, successful lunar impact (NASA)

10 April: Luna-E (1965A), lunar lander, failed to reach orbit, re-entered Earth atmosphere (USSR)

9 May: Luna 5, intended lunar lander, retro-rocket failure led to lunar impact (USSR)

8 June: Luna 6, intended lunar lander, failed to land, flew past Moon (USSR)

18 July: Zond 3, intended lunar flyby, went into heliospheric orbit (USSR)

4 October: Luna 7 (also Lunik 7), intended lunar lander, failure led to lunar impact (USSR)

12 November: Venera (3MV-4 No.4), partially successful Venus flyby, contact lost (USSR)

16 November: Venera 3, Venus lander, possible crash, no data returned (USSR)

23 November: Kosmos 96, Venus lander, failure after launch (USSR)

3 December: Luna 8, intended lunar lander, failure led to lunar impact (USSR)

16 December: Pioneer 6, successful space weather observations (NASA)

1966

31 January: Luna 9, successful lunar lander mission (USSR)

1 March: Kosmos 111, intended lunar orbiter mission, failed after launch (USSR)

31 March: Luna 10, first successful lunar orbiter (USSR)

30 April: Luna 1966A, intended lunar orbiter, failed on launch (USSR)

30 May: Surveyor 1, successful lunar lander (NASA)

1 July: Explorer 33, lunar orbiter, sent into eccentric Earth orbit (NASA)

10 August: Lunar Orbiter 1, first successful American lunar orbiter (NASA)

24 August: Luna 11, lunar mission, achieved orbit, returned some data, no images (USSR)

20 September: Surveyor 2, intended lunar lander, lost contact (NASA)

22 October: Luna 12 (Lunik 12), successful lunar orbiter (USSR)

6 November: Lunar Orbiter 2, successful lunar orbiter (NASA)

21 December: Luna 13, successful lunar lander (USSR)

1967

4 February: Lunar Orbiter 3, successful lunar orbiter (NASA)

17 April: Surveyor 3, successful lunar lander (NASA)

8 May: Lunar Orbiter 4, successful lunar orbiter/imaging (NASA)

12 June: Venera 4, successful Venus atmospheric probe (USSR)

14 June: Mariner 5, successful Venus flyby, entered heliocentric orbit (NASA)

7 June Kosmos 167, intended Venus mission, fell back to Earth 7 days post-launch (USSR)

4 July: Surveyor 4, intended lunar lander, lost contact just prior to touchdown (NASA)

9 July: Explorer 35 (IMP-E), successful interplanetary medium probe (NASA)

1 August: Lunar Orbiter 5, successful lunar orbiter (NASA)

8 September: Surveyor 5, successful lunar lander (NASA)

28 September: Zond 1967A, intended lunar test orbit, failed on launch (USSR)

7 November: Surveyor 6, successful lunar lander (NASA)

22 November: Zond 1967B, intended lunar flyby test, failed on launch (USSR)

21 December: Luna 13, successful soft landing on the Moon (USSR)

1968

7 January: Surveyor 7, successful lunar lander/imaging (NASA)

7 February: Luna 1968A, intended lunar orbiter, failed on launch (USSR)

2 March: Zond 4, test flight, destroyed on re-entry (USSR)

7 April: Luna 14 (Lunik 14), successful lunar orbiter (USSR)

23 April: Zond 1968A, intended uncrewed lunar test flight, second-stage failure (USSR)

15 September: Zond 5, successful lunar flyby/Earth return, carrying animals (USSR)

10 November: Zond 6, successful lunar flyby/Earth return, crashed due to parachute failure (USSR)

21 December: Apollo 8, successful crewed lunar orbiter (NASA)

1969

5 January: Venera 5, successful robotic Venus probe (USSR)

10 January: Venera 6, successful robotic Venus probe (USSR)

20 January: Zond A, lunar flyby/return, failed on launch (USSR)

19 February: Luna A, lunar rover, failed on launch (USSR)

21 February: Zond L1S-1, intended lunar orbiter, failed on launch (USSR)

25 February: Mariner 6, successful Mars flyby (NASA)

3 March: Apollo 9, successful third crewed test mission in Apollo series (NASA)

27 March: Mariner 7, successful Mars flyby (NASA)

2 April: Mars B, intended Mars Orbiter, failed on launch (USSR)

18 May: Apollo 10, crewed lunar mission to lunar orbit and return (NASA)

14 June: Luna C (Luna E-8-5 No.402), intended lunar sample return, failed on launch (USSR)

3 July: Zond L1S-2, intended lunar orbiter, failed on launch (USSR)

13 July: Luna 15, intended lunar orbiter/lander/sample return, crashed on lunar surface (USSR)

16 July: Apollo 11, first successful manned landing, collect samples (NASA)

7 August: Zond 7, successful lunar flyby/return to Earth (USSR)

23 September: Kosmos 300, intended lunar sample return, stranded in Earth orbit (USSR)

22 October: Kosmos 305, intended lunar sample return, stranded in Earth orbit (USSR)

14 November: Apollo 12, crewed lunar landing mission (NASA)

1970

6 February: Luna 1970A, intended lunar sample return, failed on launch (USSR)

19 February: Luna 1970B, intended lunar orbiter, failed on launch (USSR)

11 April: Apollo 13, crewed lunar mission, aborted due to explosion/returned to Earth (NASA)

17 August: Venera 7, successful Venus soft lander/data transmission (USSR)

22 August: Kosmos 359, intended Venus probe, escape stage failure (USSR)

12 September: Luna 16, successful lunar lander/sample return (USSR)

20 October: Zond 8, successful lunar flyby/return to Earth (USSR)

10 November: Luna 17/Lunokhod 1, successful lunar rover (USSR)

1971

31 January: Apollo 14, successful crewed lunar landing (NASA)

9 May: Mariner 8, Intended Mars flyby, failed on launch (NASA)

10 May: Kosmos 419, intended Mars orbiter/lander (USSR)

19 May: Mars 2, Mars orbiter/intended Lander, ended in impact (USSR)

28 May: Mars 3, Mars orbiter/lander, first soft landing on Mars, first partial image (USSR)

30 May: Mariner 9, successful Mars orbiter (NASA)

26 July: Apollo 15, successful crewed lunar landing/first human rover use on Moon (NASA)

2 September: Luna 18, intended Lunar sample return, crash landing/lost contact (USSR)

28 September: Luna 19, successful lunar orbiter (USSR)

1972

14 February: Luna 20, successful lunar sample return (USSR)

3 March: Pioneer 10 (Pioneer F), successful Jupiter flyby (NASA)

27 March: Venera 8, successful Venus probe (USSR)

31 March: Kosmos 482, intended Venus probe, failed to escape LEO/fell to Earth (USSR)

16 April: Apollo 16, successful crewed lunar landing (NASA)

23 November: Soyuz L3, intended lunar orbiter, failed on launch (USSR)

7 December: Apollo 17, last Apollo crewed lunar landing (NASA)

1973

8 January Luna 21/Lunokhod 2, lunar rover (USSR)

5 April: Pioneer 11, successful Jupiter/Saturn flyby (NASA)

14 May: Skylab, successful crewed Earth space station/Earth studies experiments (NASA)

10 June: Explorer 49 (RAE-B), lunar orbiter/radio astronomy explorer (NASA)

21 July: Mars 4, Mars flyby, intended Mars orbiter, unable to orbit Mars (USSR)

25 July: Mars 5, Mars orbiter mission, partially successful, lost contact (USSR)

5 August: Mars 6, intended Mars lander mission, contact lost on entry to Mars atmosphere (USSR)

9 August: Mars 7, Mars flyby/intended lander, missed entry, entered heliocentric orbit (USSR)

4 November: Mariner 10, Venus/Mercury flyby (NASA)

1974

2 June: Luna 22, successful lunar orbiter (USSR)

28 October: Luna 23, intended lunar sample return/damaged on landing (USSR)

10 December: Helios-A, successful solar observations (NASA/DLR)

1975

8 June: Venera 9, successful Venus orbiter/lander, sent first images from Venus (USSR)

14 June: Venera 10, successful Venus orbiter/lander (USSR)

20 August: Viking 1, successful Mars orbiter/lander (NASA)

9 September: Viking 2, successful Mars orbiter/lander (NASA)

1976

15 January: Helios-B, successful solar observing mission, closest approach to Sun (NASA/DLR)

9 August: Luna 24, intended lunar sample return, reached Moon but no return (USSR)

1977

20 August 1977: Voyager 2, successful Jupiter/Saturn/Uranus/Neptune flyby/interstellar mission (NASA)

5 September: Voyager 1, successful Jupiter/Saturn flyby, interstellar mission (NASA)

1978

20 May: Pioneer Venus 1, successful Venus orbiter (NASA)

8 August: Pioneer Venus 2, successful Venus atmospheric probe mission (NASA)

12 August: ISEE-3/ICE, successful Comet Giacobini-Zinner and Halley flybys, solar wind studies (NASA/ESA)

9 September: Venera 11, successful Venus lander mission, flight platform continued in heliocentric orbit/interplanetary space studies (USSR)

15 September: Venera 12, partially successful Venus orbiter/lander, returned some data but no images (USSR)

1981

30 October: Venera 13, successful Venus orbiter/lander (USSR)

4 November: Venera 14, successful Venus orbiter/lander (USSR)

1983

2 June: Venera 15, successful Venus orbiter/mapping (USSR)

7 June: Venera 16, successful Venus orbiter/mapping (USSR)

1984

15 December: Vega 1, deployed successful Venus lander, balloon probe/went on to 1986 Comet Halley flyby in 1986 (USSR)

21 December: Vega 2, deployed successful Venus lander, balloon probe/went on to 1986 Comet Halley flyby (USSR)

1985

7 January: Sakigake, successful 1986 Comet Halley flyby mission (ISAS)

2 July: Giotto, successful 1986 Comet Halley flyby (ESA)

18 August: Suisei (Planet-A), successful 1986 Comet Halley flyby (ISAS)

1988

7 July: Phobos 1, intended Mars orbiter/Phobos lander/hopper, failed en route (USSR)

12 July: Phobos 2, intended Mars Orbiter/Phobos lander/hopper, lost contact (USSR)

1989

4 May: Magellan, successful Venus orbiter/radar mapper (NASA)

18 October: Galileo, successful Jupiter orbiter, atmospheric probe (NASA)

1990

24 January: Hiten, successful lunar flyby/orbiter, deliberately crashed to lunar surface in 1993 (ISAS)

25 April: Hubble Space Telescope, successful Earth orbiting observatory (NASA/ESA)**

6 October: Ulysses, successful solar polar orbiter/Jupiter flyby, now decommissioned (NASA/ESA)

1991

30 August: Yokhoh, successful solar observing satellite (ISAS/NASA/PPARC)

1992

25 September 1992: Mars Observer, intended Mars orbiter, contact lost August 1993 (NASA)

1994

25 January: Clementine, lunar orbiter/Intended asteroid flyby, asteroid trip aborted (NASA/BMDO)

1 November: WIND, successful long-term study of solar wind from L1 orbit (NASA)

1995

2 December: SOHO (Solar Heliospheric Observatory), successful long-term study of solar wind (NASA/Astrium)

1996

17 February: NEAR Shoemaker, successful asteroid Eros orbiter/imager, first near-Earth asteroid flyby, orbit, landing (NASA)

7 November: Mars Global Surveyor: successful Mars orbiter/imaging/mapping (NASA)

16 November: Mars 96, intended Mars orbiter/lander, failed to leave Earth orbit (NASA)

4 December: Mars Pathfinder, successful Mars lander and Sojourner rover (NASA)

1997

25 August: ACE (Advanced Composition Explorer), successful long-term solar wind studies (NASA)

15 October: Cassini-Huygens, successful long-term Saturn orbiter, Titan lander (NASA/ESA/ASI)

24 December: AsiaSat 3/HGS-1/PAS-22, inadvertent lunar flyby (AsiaSat Ltd./Hughes)

1998

7 January: Lunar Prospector, successful lunar orbiter mission, deliberately crashed to surface to search for water (NASA)

3 July: Nozomi (Planet-B), partially successful Mars orbiter (ISAS)

24 October: Deep Space 1 (DS1), successful asteroid/comet flyby mission (NASA)

11 December: Mars Climate Orbiter, intended Mars orbiter, crashed after start of orbital insertion (NASA)

1999

3 January: Mars Polar Lander, intended Mars lander, Deep Space 2 penetrators, destroyed when craft crashed to surface (NASA)

7 February: Stardust, successful comet coma sample return mission (NASA)

2001

7 April: 2001 Mars Odyssey, longest-serving Mars orbiter (NASA)**

8 August: Genesis, first mission to return samples from the solar wind (NASA)

2002

3 July: CONTOUR (Comet Nucleus TOUR), unsuccessful mission to study comet nuclei, lost (NASA)

2003

9 May: Hayabusa (Muses-C), successful asteroid orbiter, first sample return from an asteroid (JAXA)

2 June: Mars Express, successful Mars orbiter, lander failed to deploy upon landing (ESA)**

10 June: Spirit (MER-A), successful Mars rover, extended several times, eventually went to sleep in 2010, no contact since then (NASA)

8 July: Opportunity (MER-B), still-functioning Mars rover as of mid-2017 (NASA)**

1 September: SMART 1, successful lunar orbiter, crashed into Moon at end of mission (ESA)

2004

2 March: Rosetta, successful comet orbiter, partially successful Philae lander, Rosetta hard-landed on comet at end of mission in 2016 (ESA)

3 August: MESSENGER, first successful Mercury orbiter, de-orbited on Mercury in 2015 (NASA)

2005

12 January: Deep Impact, comet rendezvous, impact/extended mission to other comets, contact lost in 2013 (NASA)

12 August 2005: Mars Reconnaissance Orbiter, successful Mars orbiter/mapping/imaging (NASA)**

9 November 2005: Venus Express, successful Venus orbiter and long-term Venus atmosphere studies (ESA)

2006

19 January: New Horizons, successful Pluto/Charon flyby, targeted to new Kuiper Belt Object flyby in January 2019 (NASA)**

22 September: Hinode (Solar-B), successful solar orbiter (JAXA/NASA/United Kingdom)**

26 October: STEREO (Solar Terrestrial Relations Observatory), partially successful twin spacecraft studying the Sun and its activity (NASA) **

2007

4 August: Phoenix, successful Small Mars Scout Lander

14 September: Kaguya (SELENE), successful lunar orbiter, de-orbited into lunar surface (NASDA)

27 September: Dawn, successful mission to orbit asteroid Vesta, then deployed to asteroid Ceres (NASA) **

24 October: Chang'e 1, lunar orbiter, deliberately crashed to Moon's surface (CNSA)

2008

22 October: Chandrayaan-1, ISRO lunar orbiter, de-orbited and crashed on lunar south pole (ISRO)

2009

7 March: Kepler, long-term extrasolar planet search (NASA)

18 June: Lunar Reconnaissance Orbiter, lunar orbiter/LCROSS impactor (NASA)**

2010

11 February: Solar Dynamics Observatory, continuous monitoring of Sun and solar activity (NASA)**

20 May: Akatsuki/Planet-C, long-term Venus orbiter (ISAS)**

15 June: PICARD, solar monitoring and helioseismology (CNES)**

1 October: Chang'e 2, lunar orbiter, sent on to asteroid Toutatis (CNSA)**

2011

5 August: Juno, Jupiter orbiter mission, arrived Jupiter 5 July 2016 (NASA)**

10 September: GRAIL, two lunar orbiters to chart lunar gravity and structure (NASA)

8 November: Phobos-Grunt, intended Martian Moon Phobos lander, failed to leave Earth orbit (USSR)

8 November: Yinghuo-1, intended Mars orbiter, lost when Phobos-Grunt burns failed (CSNA)

26 November: Curiosity Rover Mars Science Laboratory, long-running Mars rover (NASA)**

2012

30 August: Van Allen Probes, long-term study of Earth's radiation belts (NASA)

2013

27 June: IRIS (Interface Region Imaging Spectrograph), solar observations (NASA)**

6 September: LADEE (Lunar Atmosphere and Dust Environment Explorer), lunar orbiter (NASA)

14 September: Hisaki, studies of planetary atmospheres in solar system (JAXA)**

5 November: Mangalyaan, long-running Mars orbiter mission (ISRO)**

18 November: MAVEN (Mars Atmosphere and Volatile Evolution Mission), Mars atmospheric studies orbiter (NASA)**

1 December: Chang'e 3, successful lunar lander and Yutu rover (CNSA)**

2014

23 October: Chang'e 5 Test Vehicle, lunar flyby and return (CNSA)**

3 December: Hayabusa 2, asteroid mission en route to asteroid 162173 Ryugu (JAXA)**

2015

11 February: DSCOVR (Deep Space Climate Observatory), space weather and Earth observation satellite (NOAA/NASA)**

2016

14 March: ExoMars 2016, Mars orbiter/lander, lander failed, orbiter remains operational (ESA)**

8 September: OSIRIS-REx: sample return mission to asteroid 101955 Bennu (NASA)**

Future Missions (planned launch dates)

November 2017: Chang'e 5, lunar sample return mission (CNSA)

April 2018: BepiColombo, ESA and JAXA Mercury Orbiters (ESA/JAXA)

5 May 2018: InSight Mars lander (NASA)

Late 2018: Chang'e 4, reconfigured lunar far-side lander (CNSA)

2020: ExoMars 2020, Mars Rover and surface platform (CNSA/ESA/Roscosmos)

Earth Observation Satellites

(Correct as of mid-2017.)
These monitor land, ocean, atmosphere, ice sheets, interaction with the Sun, and other aspects of Earth's systems

Aqua (NASA)
AURA (NASA)
CALIPSO (NASA)
CloudSat (NASA/CSA)
Deep Space Climate Observatory (NASA)
Global Precipitation Measurement (NASA)
GRACE (NASA/DLR)
ICESat (NASA)
Jason-1 (NASA)
Jason-2 (NASA)
Jason-3 (NASA)
LAGEOS 1 and 2 (NASA)
QuikSCAT (NASA)
SEASTAR (NASA)
SORCE (NASA)
TERRA (NASA/CSA/JAXA)
TRMM (NASA/JAXA)

Weather Satellites

(Correct as of mid-2017.)
Advanced Land Observing Satellite-2 (Daichi 2) (JAXA)
Arirang-2 (KARI)
Cartosat-2A (ISRO)
CartoSat-2 (ISRO)
China-Brazil Earth Resources Satellite, CBERS-4 (CNES/INPE)
Chollian (KARI)
CryoSat-2 (ESA)
Diwata-1 (PHL-Microsat-1) DOST
DubaiSat 1 and 2 (MBRSC)
Gaofen-1 (CNSA)
Geostationary Operational. Environmental Satellites, GOES 13, 14, 15 (NASA)
Göktürk-1 (TÜBİTAK)
Göktürk-2 (TÜBİTAK)
Göktürk-3 (TÜBİTAK)
Greenhouse Gases Observing Satellite, GOSat (JAXA)
Himawari-8 (JMA)
IMS-1 (ISRO)
IRS Series (ISRO) (two currently operational)
Landsat 7 (USGS)
Landsat 8 (USGS)
NOAA-4 (NOAA/NASA)
TUBSat (LAPAN)
Megha-Tropiques (ISRO)
MetOp-A, MetOp-B (EUMETSAT)
MeteoSats 7, 8, 9, 10, 11 (EUMETSAT)
MTSAT-2/Himawari-7 (JMA)
Nigeriasat-2 (Nigeria)
NigComSat-1R (Nigeria)
Pléiades (CNES)

Polar Operational Environmental Satellites, POES (NOAA and EUMETSAT)
    NOAA-15
    NOAA-18
    NOAA-19
    METOP-B
    METOP-A
RadarSat-2 (CSA) (ESA and McDonald, Dettwiler and Associates)
Rasat (TÜBİTAK)
Resurs-P No. 1 (Roscosmos)
Resurs-P No. 2 (Roscosmos)
Sentinel 1 and 2 (ESA)
SPOT series (CNES)
Suomi-NPP (NASA)
Thaichote (GISTDA)
Thermosphere Ionosphere Mesosphere Energetics and Dynamics, TIMED (NASA)
TIROS-N 18 (NOAA/NASA) (Also known as POES)

APPENDIX G

Astronomy Satellites

Telescopes marked with ** are still in operation as of mid-2017
Note: some observatories are multi-spectral and show up in more than one list

Gamma Ray
16 November 1972: SAS-2, small astronomy satellite (NASA)
9 August 1975: Cos-B (ESA)
20 September 1979: HEAO-3 (NASA)
1 December 1979: Granat (CNRS & IKI)
1 July 1990: Gamma, (USSR/CNES/RSA)
5 April 1991: Compton Gamma Ray Observatory (NASA)
19 May 1997: Low Energy Gamma Ray Imager (INTA/Spain)
9 October 2000: High Energy Transient Explorer (NASA)
17 October 2002: International Gamma Ray Astrophysics Laboratory, INTEGRAL (ESA)**
20 November 2004: Swift Gamma-Ray Burst Mission (NASA)**
23 April 2007: Astrorivelatore Gamma ad Immagini L'Eggero (ISA)**
11 June 2008: Fermi Gamma-Ray Telescope (NASA)**
21 May 2010: Gamma-ray Burst Polarimeter (JAXA)**

Gravitational Waves
3 December 2015: LISA Pathfinder (ESA)

Infrared/Submillimetre
25 January 1983: Infrared Astronomy Satellite IRAS (NASA)
18 March 1995: Infrared Telescope in Space (ISASA/NASDA)
17 November 1995: Infrared Space Observatory, ISO (ESA)
24 April 1996: Midcourse Space Experiment, MSX (USN)
6 December 1998: Submillimeter Astronomy Satellite, SWAS (NASA)
5 March 1999: Wide Field Infrared Explorer, WIRE (NASA)
25 August 2003: Spitzer Space Telescope (NASA)**

21 February 2006: Akari (Astro F) (JAXA)
14 May 2009: Herschel Space Observatory (ESA/NASA)
14 December 2009: Wide-field Infrared Survey Explorer, WISE (NASA)

Microwave
18 November 1989: Cosmic Background Explorer, COBE (NASA)
20 February 2001: Odin (Swedish Space Corporation)**
30 June 2001: WMAP (NASA)
14 May 2009: Planck (ESA)

Orbital Particle Detectors
20 September 1975: High Energy Astrophysics Observatory 3, HEAO 3 (NASA)
3 July 1992: SAMPEX (NASA/DE)
2 June 1998: Alpha Magnetic Spectrometer 1, AMS-1, tested on shuttle *Discovery* (NASA)
15 May 2006: Payload for Antimatter Exploration and Light-nuclei Astrophysics, PAMELA (ISA/INFN/RSA/DLR/SNSB)**
19 October 2008: IBEX (NASA)**
16 May 2011: Alpha Magnetic Spectrometer 02, AMS-02, deployed on ISS (NASA)
17 December 2015: Dark Matter Particle Explorer, DAMPE (CNSA)

Radio
12 February 1997: Highly Advanced Laboratory for Communications and Astronomy, MUSES-B (ISAS)
May 2011: Spektr-R (ASC/LPI)
September 2017: Spektr-RG (RSRI)

Ultraviolet
7 December 1968: OAO-2, Stargazer (NASA)
18 April 1971: Orion 1 Space Observatory (USSR)
16 April 1972: Far Ultraviolet Camera/Spectrograph (NASA)
21 August 1972: OAO-3, Copernicus (NASA)
18 December 1973: Orion 2 Space Observatory (USSR)
30 August 1974: Astronomical Netherlands Satellite (SRON)
26 January 1978: International Ultraviolet Explorer, IUE (ESA/NASA/SERC)
23 March 1983: Astron (IKI)
24 April 1990: Hubble Space Telescope (NASA/ESA)**
2 December 1990: Broad Band X-Ray Telescope/Astro 1 (NASA)
7 June 1992: Extreme Ultraviolet Explorer, EUVE (NASA)
2 March 1993: Astro 2 (NASA)
24 June 1999: Far Ultraviolet Spectroscopic Explorer, FUSE (NASA/CNES/CSA)
13 January 2003: Cosmic Hot Interstellar Spectrometer, CHIPS microsatellite (NASA)
28 April 2003: Galexy Evolution Explorer, GALEX (NASA)
27 April 2003: Kaistsat 4 (KARI)
20 November 2004: Swift Gamma-Ray Burst Mission (NASA)**
27 June 2013: Interface Region Imaging Spectrograph, IRIS (NASA)**
14 September 2013: Hisaki, Sprint-A (JAXA)**
26 November 2013: Venus Spectral Rocket Experiment (suborbital, reusable) (NASA)
28 September 2015: Astrosat (ISRO)**

Visible
8 August 1989: Hipparcos astrometric mission (ESA)
24 April 1990: Hubble Space Telescope (NASA/ESA)**
30 June 2003: MOST (CSA)**
20 November 2004: Swift Gamma-Ray Burst Mission (NASA)**
27 December 2006: COROT (CNES/ESA)
6 March 2009: Kepler Mission (NASA)**
2013–2014: BRITE nanosat constellation (Austria/Canada/Poland)**
25 February 2013: Near Earth Object Surveillance Satellite (CSA)**
19 December 2013: Gaia astrometric mission (ESA)**
28 September 2015: Astrosat (ISRO)**

X-ray
12 December 1970: Uhuru (NASA)
30 August 1974: Astronomical Netherlands Satellite, ANS (SRON)
15 October 1974: Ariel V (SRC/NASA)
19 April 1975: Aryabhata (ISRO)
7 May 1975: SAS-C (Third Small Astronomy Satellite) (NASA)
9 August 1975: Cos-B (ESA)
6 February 1976: Cosmic Radiation Satellite, CORSA (ISAS)
12 August 1977: High Energy Astronomy Observatory, HEAO 1 (NASA)
13 November 1978: Einstein Observatory, HEAO 2 (NASA)
21 February 1979: Hakucho, CORSA-b (ISAS)
20 September 1979: High Energy Astronomy Observatory 3, HEAO 3 (NASA)
20 February 1983: Tenma, Astro-B (ISAS)
23 March 1983: Astron (IKI)
27 May 1983: EXOSAT (ESA)
5 February 1987: Ginga, Astro-C (ISAS)
1 December 1989: Granat (CNRS/IKI)
1 June 1990: Röntgen Satellite, ROSAT (NASA/DLR)
20 February 1993: Advanced Satellite for Cosmology and Astrophysics (ISAS/NASA)
25 April 1993: Array of Low Energy Imaging Sensors, Alexis (LANL)
30 December 1995: Rossi X-ray Timing Explorer, RTXE (NASA)
30 April 1996: BeppoSAX (ASI)
28 April 1999: A Broadband Imaging X-ray All-sky Survey, ABRIXAS (DLR)
23 July 1999: Chandra X-ray Observatory (NASA)**
10 December 1999: XMM-Newton (ESA)**
9 October 2000: High Energy Transient Explorer 2, HETE 2 (NASA)**
17 October 2002: International Gamma Ray Astrophysics Laboratory, INTEGRAL (ESA)**
20 November 2004: Swift Gamma Astrophysics Laboratory (NASA)**
10 July 2005: Suzaku, Astro-E2 (JAXA/NASA)
23 April 2007: AGILE (ISA)**
13 June 2012: Nuclear Spectroscopic Telescope Array, NuSTAR (NASA)**
28 September 2015: Astrosat (ISRO)**
17 February 2016: Hitomi, Astro-H (JAXA)

Near-future Telescopes
March 2018: Transiting Exoplanet Survey Satellite, TESS (NASA)
October 2018: James Webb Space Telescope (NASA/ESA/CSA)

References

Books

Aldrin, Buzz, *Magnificent Desolation: the Long Journey Home from the Moon* (New York: Bloomsbury Publishing, 2009)

Aldrin, Buzz & David, L., *Mission to Mars: My Vision for Space Exploration* (New York: National Geographic, 2015)

Allen, J. P., *Entering Space: An Astronaut's Odyssey* (New York: Stewart, Tabori & Chang, 1984)

Angelo, J., *Dictionary of Space Technology* (New York: Facts on File, 2003)

Beatty, J. K, Petersen, C. C. & Chaikin, A., *The New Solar System* (Cambridge: Cambridge University Press and Sky Publishing, 1999 (4th ed.))

Bova, B., *Mars* (New York: Bantam Spectra, 1993)

Bradbury, R. *The Martian Chronicles* (New York: Simon & Schuster, 2012 (reprint))

Burgess, C. & Doolan, K., *Fallen Astronauts* (Lincoln: University of Nebraska Press, 2016)

Burrows, W. E., *This New Ocean: The Story of the First Space Age* (New York: Modern Library, 1999)

Chaikin, A. C., *A Man on the Moon: The Voyages of the Apollo Astronauts* (London, UK: Penguin Books, 2007)

Clarke, A. C., *Man and Space* (Fairfax, VA: Life Science Library 1964)

Clarke, A. C., 'The Sentinel', *Ten Story Fantasy* (New York: Signet, 1974 (reprint))

David, L., *Mars: Our Future on the Red Planet* (New York: National Geographic Books, 2016)

Evans B. & Harland, D. M., *NASA's Voyager Missions* (New York: Springer, 2003)

Grant, G., *Flight: The Complete History of Aviation* (London: DK Publishers, 2017)

Hadfield, C. A., *An Astronaut's Guide to Life on Earth* (Boston: Little, Brown and Company, 2013)

Heinlein, R. A., *The Man Who Sold the Moon* (Riverdale: Baen Books, reissued 2000)

Heinlein, R. A., *Expanded Universe* (Riverdale, Canada: Baen Books, 2005)

Heppenheimer, T. A., *Countdown: A History of Space Flight* (Hoboken, NJ: Wiley, 1999)

Kaufman, M., *Mars Up Close: Inside the Curiosity Mission* (New York: National Geographic, 2014)

References

Landis, G. A., *Mars Crossing* (New York: Tor Books, 2000)

McEwen, A. S., Hansen-Koharcheck, C. J. & Espinoza, A., *Mars: The Pristine Beauty of the Red Planet* (Tucson: University of Arizona Press, 2017)

Michener, J. A., *Space* (New York: Random House, 1982)

National Commission on Space, *Pioneering the Space Frontier* (New York: Bantam Books, 1986)

Oberg, J. E., *Red Star in Orbit* (New York: Random House, 1981)

Oberth, H., *Die Rakete zu den Planetenräumen (The Rocket Into Planetary Space)* (1932)

Oberth, H., *Menschen im Weltraum (Man into Space)* (1953)

O'Neill, G., *The High Frontier: Human Colonies in Space* (Burlington, Canada: Apogee Books, 2000)

Petersen, C. C., 'The Media Treatment of the Hubble Space Telescope' (thesis) (Boulder: University of Colorado, 1996)

Petersen, C. C. & Brandt, J. C., *Hubble Vision: Further Adventures with the Hubble Space Telescope* (Cambridge: Cambridge University Press, 1998)

Petersen, C. C., 'The Birth and Evolution of the Planetarium', *Information Handling in Astronomy: Historical Vistas*, ed. A. Heck (Netherlands: Kluwer Academic Publishers, 2003)

Petersen, C. C. & Brandt, J. C., *Visions of the Cosmos* (Cambridge: Cambridge University Press, 2003)

Petersen, C. C. & Petersen, M. C., 'The Birth and Evolution of the Planetarium', *Information Handling Systems in Astronomy: Historical Vistas*, ed. A Heck (Netherlands: Kluwer Academic Publishers, 2003)

Petersen, C. C., 'The International Planetarium Society', *Organizations and Strategies in Astronomy*, 6, ed. A. Heck (New York: Springer, 2006)

Petersen, C. C., *Astronomy 101: From the Sun and Moon to Wormholes and Warp Drive* (Blue Ash, OH and New York: Adams Media, 2013)

Sherr, L., *Sally Ride: America's First Woman in Space* (New York: Simon & Schuster, 2015)

Silverberg, R., *Ringworld* (New York: DelRay Books/Random House, 1985)

Spitzer, Lyman Jr, 'Report to Project Rand: Astronomical Advantages of an Extra-Terrestrial Observatory', reprinted in *NASA SP-2001-4407: Exploring the Unknown*, ch. 3 pt 1, p. 546.

Squyres, S., *Roving Mars* (New York: Hyperion Books, 2005)

Swift, D. W., *Voyager Tales* (Reston, VA: AIAA, 1997)

Tsiolkovsky, K. E., 'The Exploration of Cosmic Space by Means of Reaction Devices', *Science Review*, 5 (1903)

United Press International, *Gemini: America's Historic Walk in Space* (New Jersey: Prentice-Hall Books, 1965)

Von Braun, W. & White, H. J., *The Mars Project* (Champaign: University of Illinois Press, 1962)

Von Braun, W., *The History of Rocket Technology* (Detroit: Wayne State University Press, 1964)

Webber, D., *No Bucks, No Buck Rogers: the Business of Commercial Space* (Boca Raton, FL: Universal Publishers, 2017)

Winter, F., *Prelude to the Space Age: The Rocket Societies, 1924-1940* (Washington, DC: Smithsonian Press, 1983)

Zubrin, R., *The Case for Mars* (Florence, MA: Free Press, 2011)

Articles and Papers

'Addressing India's Strategic Needs Through a National Space Law' [https://jsis.washington.edu/news/addressing-indias-strategic-needs-national-space-law/]

'A Little-known Story from the Life of John Glenn' [http://www.cbsnews.com/news/a-little-known-story-from-the-life-of-john-glenn/]

'A Method of Reaching Extreme Altitudes' (Robert H. Goddard) [http://www2.clarku.edu/research/archives/pdf/ext_altitudes.pdf]

'Alexi Leonov: The First Man to Walk in Space' [https://www.theguardian.com/science/2015/may/09/alexei-leonov-first-man-to-walk-in-space-soviet-cosmonaut]

Apollo Fact Sheets (NASA) [https://www.nasa.gov/centers/langley/news/factsheets/Apollo.html]

Astronaut Samantha Cristoforetti [http://samanthacristoforetti.esa.int/]

'Basics of Space Flight' [https://solarsystem.nasa.gov/basics/]

'Benefits of Space: Agriculture' [http://www.unoosa.org/oosa/en/benefits-of-space/agriculture.html]

Biography of Neil Armstrong [https://www.nasa.gov/centers/glenn/about/bios/neilabio.html]

'China Has Completed its Space Station' [http://gbtimes.com/china/china-has-completed-its-space-station-core-module-ready-launch-2018]

Chronicle of Soviet-Russian Space Program [http://en.roscosmos.ru/174/]

Communications Satellites: Making the Global Village Possible [https://history.nasa.gov/printFriendly/satcomhistory.html]

CU Students at LASP [http://lasp.colorado.edu/home/about/cu-students-at-lasp/]

'Earth-observing Companies Push for More-advanced Science Satellites' [https://www.nature.com/news/earth-observing-companies-push-for-more-advanced-science-satellites-1.22034]

Exploring the Unknown, Selected Documents in the History of the US Civil Space Program [https://history.nasa.gov/SP-4407/vol3/cover.pdf]

'First Woman in Space Ready for No-Return Mission to Mars' [https://sputniknews.com/world/20130607181561637-First-Woman-in-Space-Ready-for-No-Return-Mission-to-Mars/]

Manned Orbiting Laboratory Declassified [http://www.nro.gov/foia/declass/MOL.html]

Morehead Planetarium History [http://moreheadplanetarium.org/about/history]

NASA Historical Data Book Volume II Programs and Projects 1958-1968 [https://ntrs.nasa.gov/archive/nasa/casi.ntrs.nasa.gov/19880016046.pdf]

NASA Historical Data Book Volume III Programs and Projects 1969-1978 [https://ntrs.nasa.gov/archive/nasa/casi.ntrs.nasa.gov/19880016046.pdf]

NASA's Resource Prospector [https://www.nasa.gov/resource-prospector]

'Neil deGrasse Tyson: Don't Leave Space Exploration Up to Private Companies' [http://bgr.com/2015/12/03/neil-degrasse-tyson-interview-space-exploration/]

'New Law Aims to Expand Japan's Space Business' [http://www.nippon.com/en/currents/d00294/]

Optical History, Lyman Spitzer: Large Space Telescope [http://ecuip.lib.uchicago.edu/multiwavelength-astronomy/optical/history/11.html]

'Public-private partnerships helping to take public agendas forward' [http://www.unesco.org/new/en/media-services/single-view/news/publicprivate_partnerships_helping_to_take_public_agendas/]

'Robert H. Goddard: American Rocket Pioneer' [https://siarchives.si.edu/history/exhibits/stories/robert-h-goddard-american-rocket-pioneer]

'Russia Approves its 10-year space Strategy' [http://www.planetary.org/blogs/guest-blogs/2016/0323-russia-space-budget.html]

'Russian Approves Long-awaited 10-year Space Budget' [https://themoscowtimes.com/articles/russia-approves-long-awaited-10-year-space-budget-52194]

'Science Education in the United States: The Good, the Bad, and the Ugly' [https://history.nasa.gov/sp4801-chapter21.pdf]

'Sputnik and Science Education' [http://www.nas.edu/sputnik/ruther1.htm]

State of the Dome, 2014 [https://www.lochnessproductions.com/reference/2014state/2014stateofthedome.html]

'Study Investigates How Men and Women Adapt Differently to Space Flight' [https://www.nasa.gov/content/men-women-spaceflight-adaptation/]

'The new space race' [http://www.bbc.co.uk/news/resources/idt-sh/disruptors_the_new_space_race]

Valentina Tereshkova [https://airandspace.si.edu/people/historical-figure/valentina-tereshkova]

'What It Feels Like to Ride the Shuttle' [http://www.businessinsider.com/what-it-feels-like-to-ride-the-shuttle-2014-11]

'Yuri Gagarin's First Speech about His Flight to Space' [https://www.theatlantic.com/technology/archive/2011/04/yuri-gagarins-first-speech-about-his-flight-into-space/237134/]

'Mars Direct: A Simple, Robust, and Cost Effective Architecture for the Space Exploration Initiative', Zubrin, et al. [http://www.marspapers.org/paper/Zubrin_1991.pdf]

Selected Websites

Aerojet Rocketdyne [http://www.rocket.com/]

Arianespace [http://www.arianespace.com/]

Association of Science-Technology Centers [http://www.astc.org/]

Aviation Week [http://aviationweek.com/]

B612 Foundation [https://b612foundation.org/]

Ball Aerospace [http://www.ball.com/aerospace]

Bigelow Aerospace [https://bigelowaerospace.com/]

Blue Origin [https://www.blueorigin.com/]

Boeing Starliner [http://www.boeing.com/space/starliner/]

Cassini Mission [https://saturn.jpl.nasa.gov/]

Celestis [https://www.celestis.com/]

Challenger Centers [https://www.challenger.org/]

Chandra X-ray Observatory [http://chandra.harvard.edu/]

China Manned Space [http://en.cmse.gov.cn/]

ComSat [http://www.comsat.com/]

Deep Space Industries [https://deepspaceindustries.com/]

DLR [https://dlr.de]

ESA Spinoffs [http://www.esa.int/Our_Activities/Space_Science/Spin-off_technologies]

European Space Agency [https://esa.int]

FUNCube: UK Amateur Radio Educational Satellite [http://warehouse.funcube.org.uk/]

Hubble Space Telescope [https://hubblesite.org]

Indian Space Research Organization [http://www.isro.gov.in/]
Inmarsat [http://www.inmarsat.com/]
Intelsat [http://www.intelsat.com/]
International Planetarium Society [http://www.ips-planetarium.org/]
Italian Space Agency [http://www.asi.it/en]
James Webb Space Telescope [http://webbtelescope.org/]
Japanese Aerospace Exploration Agency [http://global.jaxa.jp/]
Kepler Mission [https://www.nasa.gov/mission_pages/kepler/main/index.html]
Leonard David's Inside Outer Space [https://www.leonarddavid.com/]
Loch Ness Productions [https://www.lochnessproductions.com]
Lockheed Martin [http://www.lockheedmartin.com/us.html]
Mars Exploration Pages (NASA) [https://mars.jpl.nasa.gov/]
Mars Orbiter Mission [http://www.isro.gov.in/pslv-c25-mars-orbiter-mission]
Mars Society [http://www.marssociety.org/]
Max Planck Institutes [https://www.mpg.de/institutes]
Moon Express [http://www.moonexpress.com/]
NASA Main Pages [https://nasa.gov]
NASA History Program Office [https://history.nasa.gov/naca/]
NASA Spinoffs [https://spinoff.nasa.gov/]
National Space Society [http://www.nss.org/]
Orbital ATK [https://www.orbitalatk.com/]
Planetary Defense Coordination Office (NASA) [https://www.nasa.gov/
    planetarydefense/overview#neo]
Planetary Resources [http://www.planetaryresources.com/#home-intro]
Planetary Science Institute [https://www.psi.edu/]
Planetary Society [http://planetary.org/]
RocketLab [https://rocketlabusa.com/]
Roscosmos (Russian Space Agency) [http://en.roscosmos.ru/]
Rosetta Mission [http://rosetta.esa.int/]
Russian Space News [http://russianspacenews.com/]
Russian Space Web [http://www.russianspaceweb.com/]
Scaled Composites [http://www.scaled.com/]
Secure World Foundation [https://swfound.org/]
SkyLab (NASA) [https://www.nasa.gov/mission_pages/skylab]
Southwest Research Institute (SWRI) [http://www.swri.org/]
Space Adventures [http://www.spaceadventures.com/]
Space.com [https://www.space.com/]
Space Daily [http://www.spacedaily.com/]
Space Flight Insider [http://www.spaceflightinsider.com/]
Space News [http://spacenews.com/]
SpaceOps Australia [http://spaceops.com.au/]
SpaceX [http://www.spacex.com/]
STEREO Mission (NASA) [https://stereo.gsfc.nasa.gov/]
TheSpacewriter [https://www.thespacewriter.com]
UK Space Agency [https://www.gov.uk/government/organisations/uk-space-agency]
United Launch Alliance [http://www.ulalaunch.com/]
United Nations Office for Outer Space Affairs [http://www.unoosa.org/]
Virgin Galactic [http://www.virgingalactic.com/]
World View Enterprises [https://worldview.space/]
XMM-Newton Mission [https://www.cosmos.esa.int/web/xmm-newton]
ZeroG [http://www.gozerog.com/]